TO RULE
THE NIGHT

To Bryan College

my Best Wishes For

Your Life in Him

His Love From The Moon

John Irwin
Apollo 15 ☾

22 July 1982

(On the occasion of their visit
to the 1982 Summer Bible Conference,
at which both Colonel and Mrs. Irwin
spoke.)

TO RULE
THE NIGHT

The Discovery Voyage
of Astronaut Jim Irwin

by JAMES B. IRWIN
with WILLIAM A. EMERSON, JR.

A. J. HOLMAN COMPANY
Division of J. B. Lippincott Company
Philadelphia & New York

48098

U. S. Library of Congress Cataloging in Publication Data

Irwin, James Benson.
  To rule the night.

  1. Irwin, James Benson. 2. Project Apollo.
3. Religion and astronautics. I. Emerson, William
A., birth date   joint author. II. Title.
TL789.85.I78A3   629.4′092′4 [B]   73-11410
ISBN-0-87981-024-6

To Mary and our five J's; Joy, Jill, Jimmy, Jan, and now Joe Chan from Vietnam. I am indebted to my family for their confidence in my flight and for keeping me humble and down to earth on my return. The first question the children asked when I got back was, "Daddy, when are you going back to the moon?"

I am left with a prayer of thanks that the whole family can now be with me on the High Flight as witnesses to the love of Jesus Christ.

CONTENTS

PROLOGUE   11

1 BLAST-OFF   25

2 ON THE MOON   60

3 FLYING HOME   89

4 REUNION   111

5 EARLY YEARS   125

6 MILITARY LIFE   146

7 GROUNDED   179

8 ASTRONAUT TRAINING   197

9 HIGH FLIGHT   223

EPILOGUE   245

A section of photographs follows
  the Epilogue.

7

TO RULE
THE NIGHT

PROLOGUE

WHEN YOU LEAN far back and look up, you can see the
earth like a beautiful, fragile Christmas tree ornament hanging against the blackness of space. It's as if you could reach
out and hold it in your hand. That's a feeling, a perception, I
had never anticipated. And I don't think it's blasphemous for
me to say I felt I was seeing the earth with the eyes of God. I
believe, looking back on it now, the good Lord did have His
hand in it. For me to travel such a roundabout way, and
finally end up in the space program, and then go to the moon
—it's amazing it ever happened.

I always tell people that before the flight of Apollo 15 I
considered myself a technician, a test pilot, the operator of a
spacecraft—really a nuts-and-bolts type. And I didn't have an
unblemished record, either. I had made many mistakes,
goofed up many times, had my ups and downs, physically and
spiritually. Then, on my third try, just as I reached the age
limit, I was accepted into the astronaut program. This was
the most elite group in the service, and a great honor, but
even then I had no idea that I would ever be chosen to
actually go to the moon.

When I was selected as an astronaut, it was something
of a miracle that I was still in the Air Force. I had squeaked
into Annapolis with a fraction of a point to spare on the

substantiating examination. Then, before I had even graduated from the Naval Academy, I was ready to give it up. The Navy was so outmoded, so old-fashioned in its policies, that I didn't want any part of that branch of the service. It seemed ridiculous to spend so much time at sea, for example. Sure, it was nice to go for a short boat ride, but why spend six months or a year away from your family? There were so many more interesting things to do on land—or even in the air.

It was providential that the Air Force was created in 1949, and that they were looking for young officers. The Air Force was able to take 20 percent of our class at Annapolis, and fortunately I drew a low number. If I had been stuck in the Navy, I probably would have served the minimum time and then resigned.

As it turned out, I had some fantastic opportunities in the Air Force. I was even assigned to be the first and only test pilot on the world's highest- and fastest-flying airplane, the YF-12A, which made me so proud I thought I was the hottest test pilot in the sky. But truthfully, throughout most of my Air Force career I was sorry I hadn't gotten out of the service back in 1955 and gone with a commercial airline. I was sure I could have had a more satisfying life if I had flown as an airline pilot until I was sixty-five and then just retired. I couldn't think of a more relaxing life, particularly for a man like me who loves to fly. I would make enough money and still have a lot of free time.

I can't imagine life without flying now, but it wasn't any instant love affair. During my pilot training in Hondo, Texas, I came to the conclusion that I didn't care much for it. I had had about ten hours when I decided I wasn't cut out to be a pilot. Aviation wasn't challenging or exciting. I was uncomfortable in that bumpy Texas air. It didn't make me airsick, but I didn't feel comfortable.

I told my instructor, "I don't think this is right for me. I've sort of lost interest."

"Jim," he said, "if you feel that way, why don't you go over and see the Base Commander and tell him about it?"

The Base Commander's name was Colonel Irvin; I went to see him and explained to him that I really did appreciate what they were trying to do down there but that I wasn't that interested in flying and wanted to resign from flight training. He took it as a personal affront.

"I can't have that, Irwin. I can't tolerate that here," he said. He sent me to talk to the Commandant of Students.

The Commandant of Students was a tall, heavy man, and none of the guys really liked him. He was bad news. I saluted. "Lieutenant Irwin reporting, sir."

"Get out of here, Irwin," he said. "Come back in and report properly."

I got out and stayed in the hall until he came for me. Then I reported again and told him that I didn't particularly care for the outfit and wanted to leave.

He came unglued. "The only way you'll get out of here, Irwin, is to sign a statement that you're afraid to fly." I refused to do that. "Then if you are not afraid to fly, you had better go back to flying." I did.

At that point, I don't think I had any outstanding aptitude for flying. I was probably average. But I was relaxed and I slept a whole lot. Half the guys in my little flight group washed out. I've always thought that the only reason I didn't wash out was because I didn't clutch up. Maybe the fact that I wasn't particularly interested in the thing made it easier for me to go through with it, that and the fact that I had a great little guy from south Texas named Ed Siers as my flight instructor. Ed probably had fifteen to twenty thousand hours at the time; he was an old crop-duster and had done all types of flying. It was Ed who sent me to the Base Commander to resign from flight training, and when I told him about the response I had gotten he accepted it.

"We'll try to work this thing out," he said. "I'll try to find some smooth air for you."

Ed didn't find much smooth air, but he took me through my training. When I was knocked out with pneumonia and lost two weeks, he even volunteered to fly with me on weekends and helped me catch up. So I stuck with it, and soloed about average, and graduated with my class.

As much as I liked Ed, I discovered something new about myself when I soloed. When I could get into an airplane by myself, with no instructor along, it was a great relief. It was good to be rid of all that noise in the back seat. When I could get up there in the sky, close to God, all by myself, I flew better. That solitude was and is complete joy for me.

After I had gotten my wings, I had an experience that captured me. I had picked an assignment at the Air Force base at Yuma, Arizona, because it **was** near the mountains, so I could ski and climb, and because **it** was near my folks on the West Coast. (At that point in my life, I was interested in having the Air Force join me.) When I got to Yuma I saw P-51s there on the runway. Just seeing them was all it took.

The 51 has a tremendous engine packed into a long pointed nose that angles up so you can't see the runway and have to do S turns when you taxi out in it. I'll never forget the thrill of flying that bird. It had about 60 inches of manifold pressure, which is twice the pressure of anything I had flown up to that point. On my first takeoff, I threw the power to it and could hear the roar. You really feel you are part of the engine. I was 3,000 feet in the air before I even remembered to pull the landing gear up. Felt like I was going straight up. And I was behind the airplane, probably for the first time in my life; usually I am thinking way ahead of the airplane. But that 51 really caught me by surprise. It opened up a whole new thrill for me in flying. Suddenly I came into my own. All the tedium and boredom and everything else I had been trying to escape was behind me. I was hooked.

From that time on I lived to fly the hottest stuff that the Air Force could build. And I flew it. It got to the point that they always knew when Jim Irwin took off, because I would go right straight up in the most vertical way, on an angle that was a hairline from stalling. If that engine had conked out I would probably have had it. I logged thousands and thousands of hours, flying every moment I could get my hands on a plane. Even when I was towing targets I would take everybody's shift and sometimes fly eight hours a day. I scrambled for every assignment, every training course that would give me a chance at the newest jet or rocket plane the Air Force was putting out.

But in those years my career was forever being blocked, and I felt that things weren't working out for me. I was grounded many times for violations—a little too relaxed and high-spirited, maybe. And then I was almost wiped out in a terrible air accident. I thought I'd lost my chance to fly the YF-12A, that test pilot's dream. When I had recovered sufficiently to report for duty, they told me, "No, Captain Irwin. You are a pilot and you have had a concussion. You have even had amnesia. We cannot let you fly for at least a year while we completely evaluate you."

In 1963 I applied to NASA's space program, but they turned me down. I believe they felt the accident was too recent. The second time around, NASA was looking for scientists with doctorates and I couldn't qualify. I was hounded by the history of my injury and fast approaching the age limit for astronauts. In 1966 I made one last, despairing effort.

As I look back, I realize I didn't have the purest motives in the world. I wanted the obvious advantages that the program offered. There was a lot of additional income—why, we'd heard that the astronauts were getting a tremendous amount of money, something like $15,000 a year extra. And there was the chance to do some great and interesting flying and to do something worthwhile. The possibility of becoming famous seemed remote, but it didn't discourage me. Maybe I could afford to put my kids through college. As far as I was concerned, it was just the space program. It wasn't commonly known that they were picking guys to go to the moon. I thought there might be one or two moon shots, but I didn't have any idea what the scheduling of the Apollo program would be. I thought there might be a chance to go into orbit some day—that was the greatest opportunity.

My boss at the Air Defense Command in Colorado Springs, Col. Wilton Earle, went to bat for me. He must have contacted all the generals he had ever known in the Air Force. I don't know what he told them, but obviously it had a powerful effect. It was fortunate for me that this time around, in the selection of the fifth group, they picked nineteen, the largest number ever. There were examinations and interviews and physicals and all the rest of it, and finally the announcements were made. Jim Irwin was selected.

This was the highest honor I could imagine. I had only one way of explaining this mind-boggling thing that had happened to me, and it didn't come to me until the flight of Apollo 15 was over and I had a chance to reflect on it. The

PROLOGUE

Lord wanted me to go to the moon so I could come back and do something more important with my life than fly airplanes. During the years of training and the many months of competition with the other astronauts for assignment as a primary crewman, I had been so absorbed in preparing for the scientific flight that it never even occurred to me how high the spiritual flight could be.

As we reached out in a physical way to the heavens, we were moved spiritually. As we flew into space we had a new sense of ourselves, of the earth, and of the nearness of God. We were outside of ordinary reality; I sensed the beginning of some sort of deep change taking place inside of me. Looking back at that spaceship we call earth, I was touched by a desire to convince man that he has a unique place to live, that he is a unique creature, and that he must learn to live with his neighbors.

We never felt that we were moving after we left earth orbit. We felt that we were stationary. There went the earth, moving out away from us. And all of a sudden, there was the moon; it started coming in to us. Maybe we needed this to give us a sense of security. It may have comforted us to feel that we were stable and these other bodies were moving.

The first time we could see the whole earth, we saw it as a ball in the sky. It was about the size of a basketball, and the most beautiful thing you could ever want to see in all your life. Then, as we got farther and farther away, it diminished in size. We saw it shrink to the size of a baseball, and then to the size of a golf ball, and finally to the size of a marble. From the moon the earth looked just like a marble, the most beautiful marble you can imagine. The earth is uncommonly lovely. It is the only warm living object that we saw in space on our flight to the moon.

I wish I had been a writer or a poet, so that I could

convey more adequately the feeling of this flight. It is a great challenge to a man to try to help people feel the same thing that he felt. I know that others will not have the opportunity to visit the moon, but yet because of them I was able to go. I went for everyone—not only the American taxpayers but the rest of the people on earth. Indirectly, everyone on earth was a part of this flight. It was a human effort, and all human beings can feel proud that another human being made a trip to the moon and came back to earth. I feel the responsibility of being a representative, and I have spent almost all my time since I came back giving my account of the flight of Apollo 15, talking to people directly, over radio and television, and now through this book.

During this sort of flight, you are too busy to reflect on the splendor of space or on the secret awakenings that come from the inner flight that takes place at the same time. You have to try to register these experiences and examine them later. It has been sort of a slow-breaking revelation for me. The ultimate effect has been to deepen and strengthen all the religious insight I ever had. It has remade my faith. I had become a skeptic about getting guidance from God, and I know that I had lost the feeling of His nearness. On the moon the total picture of the power of God and His Son Jesus Christ became abundantly clear to me.

I felt an overwhelming sense of the presence of God on the moon. I felt His spirit more closely than I have ever felt it on the earth, right there beside me—it was amazing. I didn't change my habits. I prayed at the same times that I do on earth, a brief prayer before I go to sleep and then when I wake up. But through those days there was a gradually enhanced feeling of God's nearness. And when we were struggling with the difficult tasks on the first EVA (Extra-Vehicular Activity, i.e., activities away from our Lunar

Module on the surface of the moon), when a key string broke and I couldn't get the science station up, I prayed. Immediately I had the answer.

It was almost like a revelation. God was telling me what to do. I never asked Houston because I knew there would be a delay. I didn't have time for Houston to get an answer to me; I needed an immediate answer. I could see several logical ways to go about solving these mechanical problems, but I wanted to know the best way. I prayed, and immediately I knew the answer. I am not talking about some vague sense of direction. There was this supernatural sensation of His presence. If I needed Him I could call on Him, call on His power.

I am not the only astronaut to be affected by this experience on the moon and in space. All of us have been, but we express it in different ways. Everybody felt that they were much more efficient in space than on the earth, that they had achieved a feeling of mental power. We all thought with a new clarity, almost a clairvoyance. I could almost anticipate what Dave Scott was going to say, and I felt that I knew what he was thinking.

Ed Mitchell was the Lunar Module Pilot on Apollo 14, and he and I have talked freely about the things we felt on the moon. He would be the first to admit that he experienced this new clairvoyance, this closeness of some power that is impossible to describe. As a result, Ed has founded the Mind Science Foundation for the purpose of pursuing scientific explanations of the presence of God. I wish him all the success in the world, but to me science *is* the uncovering of God's basic laws.

I think there are things that God does not intend man to understand, things that man is to take on faith. Of course, religion is a matter of faith, and I think God means it that way. Man needs something or, better, someone to believe in, and nothing will ever surpass the beauty and simplicity of the message of Christ. I hate to see people confused about it. If you could just convince them to take it on faith and live the Christian life by faith.

It is not an accident that the lives of the Lunar Module Pilots have been more changed by the Apollo flights than the lives of the Commanders or the Command Module Pilots. The people in my slot were sort of tourists on these flights. They monitored systems that were, for the most part, not associated with control of the vehicle, so they had more time to look out the windows, to register what they saw and felt, and to absorb it. Buzz Aldrin, for instance, had a tremendous reaction; he had a nervous breakdown when he got back from the moon. He has described in his own book how the flight turned his life upside down.

The impact comes not so much from the flight itself as from the relentless publicity and exposure that the flight generates. It is a great shock getting back and becoming an instant celebrity, a hero, a sort of superman. My wife, Mary, whom I love dearly, sees this transformation in a very interesting light—not the way you would expect. I know that she feels this book is important, and I have discussed all aspects of it with her so that I could understand the experience more clearly myself. When I ask Mary, "What do you want the book to do?" she says firmly, "I want it to break down a myth in people's eyes that astronauts are infallible. They are not gods. They are human beings. They are part of families; they have wives and children. They have the same emotions and the same needs and the same feelings as everyone else."

I tell her not to worry; there is no way on earth this book could prove anything else. We have plenty of flaws to work with here, for all practical purposes. But nevertheless, even if you stay in contact with yourself, it is still hard to keep your balance when you get back. You find yourself having lunch with kings, making speeches to heads of state, riding in ticker-tape parades, and serving as a goodwill ambassador for the United States. This is a highly charged and frenetically active interval, not the best time for sorting one's image out or discovering what the experience has meant.

At a news conference in Houston, five days after we had splashed down, I was asked about this. "Frankly," I said to the reporters, "I don't think I have changed any. I'm still the same guy." But I did tell them that the beauty of the mountains of the moon had moved me, and that I had felt the presence of God.

It took me about a month to discover what had happened inside me. When I first talked to people about Apollo 15, I would tell them anything they wanted to know about space, about the scientific side. The scientific voyage of discovery was what we had spent years preparing for; on our return, this was what NASA had debriefed us on. Shortly after the flight, we started our visits around the world, goodwill missions directed by the President. Al and Dave and I participated as a crew on these occasions. They were designed to be scientific exchanges, so I didn't have the satisfaction of telling the complete story. I wasn't free to talk about my own religious experience.

However, when we visited Italy there were many questions at the press conference about the religious implications. Since Dave and Al considered me the preacher, I answered them. More and more since then, I have had the great satisfaction of being able to share the complete message. I started working unofficially on weekends, trying not to create a problem for NASA. I received so much encouragement from the churches where I had been speaking, and so much encourage-

21

ment from Mary and the children to go ahead with this work, that there has been a growing conviction about my commitment and a strengthening of my message. I feel a terrific compulsion to hit as hard as I can while I can be useful, before my fame fades.

The response from people everywhere, all nationalities and even all religions, has been tremendously moving. Everybody wants to talk to a man who has been to the moon. They think that since he has seen something they have not seen and will never see, he must know something they do not know. They are interested in the scientific voyage, but they are also interested in the mythic voyage. They are interested in what happened inside us, in our hearts and souls. They can't go to the moon, but they can take this flight.

Of course, I think people are hungering for mystery, and the moon has been a place of mystery for man as long as the human race has existed. Some people conceive of the moon as a magical and romantic object in the sky; others think of it as a holy place. Actually, it has all these elements as far as I am concerned. The moon has a powerful force; it seems to affect the feelings and the behavior of everybody. I cannot imagine a holier place.

I think of it as a very friendly place, too. When I am flying across the Rockies in a light plane and the moon rises in front of me, I look at it with fond attachment. There is the place I spent my vacation that summer of 1971: my favorite resort.

When I came back from the flight, I was baptized at the Nassau Bay Baptist Church in Houston with my daughter Jill. I had accepted Christ when I was a boy of eleven in New Port Richey, Florida, at a revival meeting, but I didn't stay as close to the Lord as I should have. I drifted away, coming back occasionally and then straying again. But after

22

the flight the power of God was working in me, and I was possessed by a growing feeling that God did have a new mission for me. I know that a flight to the moon doesn't satisfy a man's soul; he is still looking for a meaning to life, for a plan. When I reach out to people, I find they are searching too; they are looking for help.

I tell people that God has a plan for *them*. I say that if God controls the universe with such infinite precision, controlling all the motion of the planets and the stars, this is the working out of a perfect plan for outer space. I believe that He has the perfect plan for the inner space of man, the spirit of man. This plan was manifest when He sent His son Jesus Christ to die for us, to forgive us our sins, and to show us He has a plan for our lives.

It seems plain to me that the hand of God has been in my life as far back as I am able to remember. I think Providence has been a factor in every important thing that has ever happened to me. As strange as it sounds, my flight on Apollo 15 was the fulfillment of a dream I had all my life. I have talked of wanting to go to the moon since I was a young kid. My mother says that she remembers this, and some old neighbors of ours whom she talked to recently also remembered that when I was a little boy I used to point up to the moon and say, "I'm going to go up there some day."

I probably said this, but I don't know how much confidence I had that I would be able to do it. Reverses along the way made it seem unlikely that I was headed in this direction. When it happened, I felt that I was doing something I had always wanted to do. But the most startling thing to me now is not that I have made the flight but what the flight has done to me, not just spiritually but in the way I feel about the world.

Since I've been back on earth, I feel at home. No matter

where I am on earth, I feel completely at home, relaxed. I do not feel foreign, I do not feel alien. I have visited a great many countries around the earth since my return, and I look forward to visiting many more. The experience has literally made me feel a close kinship with everyone. When you see the earth from the perspective of space, you don't see any evidence of the existence of man at all. The human problems do not seem overwhelming, they seem insignificant, puny. All you see is the beauty of the land and the water.

On the mountains of the moon I had an opportunity to quote a favorite Psalm: "I will lift up mine eyes unto the hills, from whence cometh my help." As I quoted, I had the impulse to add, "But of course we get quite a bit from Houston, too." This incredible flight that we made with help from Houston is one of the extraordinary technological achievements in the history of man. I can believe it only because I understand it and can retrace it step by step. As I tell this story of the scientific voyage of Apollo 15, I will also try to tell the story of that other voyage I made with the help of God.

I

BLAST-OFF

THEY WOKE US UP at 4:30 A.M. on Monday morning, July 26, 1971. We were sleeping in the crew quarters at Cape Kennedy in windowless little rooms along a corridor on the third floor. I always felt I was in jail when I had to sleep in the crew quarters—I would much rather sleep in the beach house, where I could be completely alone, hear the ocean, and go out in the early morning for a dip. Deke Slayton, he's kind of our chief, said, "Okay, guys, this is it," and we went immediately. We didn't know whether or not it was a good day. In the crew quarters it was hard to tell.

We walked a hundred feet down the hall to the medical facility, where we were given a brief physical. We were naked except for our bathrobes, so we dropped them off for the abbreviated examination. The medical people had been watching us every day, but they just wanted to be certain we had not developed colds or infections or anything else that might have come up during the night. Dr. Jack Teagen checked us out matter-of-factly, very concentrated, with no small talk. Everything was simplified to "get on with it."

Then we went in the dining room for breakfast. Besides our crew—Dave Scott, Al Worden, and I—there were the three guys in our support crew and the three backup guys and Deke, who was always with us. We had a big steak for break-

25

fast, scrambled eggs, milk—more breakfast than we normally would have. Everyone was sort of self-contained, thinking of what was in store, that this was the real thing. As I remember, it wasn't nearly as relaxed as other breakfasts that we had down there.

Then, everyone said good-bye. They shook hands with us, and there was maintenance conversation, but we were tuning out everything extraneous. What was going on outside was an interruption. We wanted to cut it as short as possible, get down to the suit room, and get the show on the road.

I was usually the first down to the suit room, and I walked the hundred feet and there was Dr. Teagen and another doctor who looked at me as I came in. Dave and Al got there too. We stripped and they checked us over again. They applied the Bio-Medical Sensors, two to the side of the chest and two to the front for respiration and heart rate. They fastened a signal-conditioner package around our waists, like a money belt, and they attached the urine collection devices.

Everything has a special clarity. You are in a "clean room" with a surgical atmosphere. The staff people have on white garments and white hats, and there are testing devices along the wall that will be used to check the pressure and integrity of your suit. But, first, you put on your long underwear; it's plenty big. It's lightweight, has feet, and eases the task of sliding into the space suit.

After you leave the little dressing area, you walk down an aisle to a row of reclining lounge chairs. Your suit is laid out on a nearby table. The empty suit has almost as much shape as when you're in it, except for the chest area, which is partially collapsed. You sit on a lounger in your long underwear while waiting for the scheduled time to get into the suit. Everything is very carefully programmed, and at the right moment they insert you into the suit and put on the helmet and the gloves. And then they go ahead and pressurize the suit to confirm that the pressure integrity is ensured.

Your suit technician pulls a hose out of a console on the wall and pumps the suit up; you're on 100 percent oxygen. You have to prebreathe for at least three hours before launch to get rid of the excess nitrogen in your bloodstream so you don't get the bends when you start to ascend. They build up the suit to 4 pounds per square inch (psi) above sea level, which is 15 psi, giving you an environment of 19 psi. The pure oxygen smells kind of good. This 19 psi approximates the pressure you will have in your suit when you walk on the moon. Then they let the suit down to sea level, and you can lie on your couch and continue prebreathing with the hoses still connected.

I had them put a towel over my helmet to keep out the light. I always tried to get as much sleep as I could, and I was having a few calm thoughts as I dozed off. The day before had been exhausting, physically and emotionally. I was physically tired because I had played tennis until I had completely dehydrated myself; I had to drink three bottles of Gatorade. Later I discovered I had probably depleted my potassium level and affected my heart by wearing myself out. And I was emotionally drained from saying good-bye to my family. Since we were quarantined, they had been behind glass and I talked to them by microphone. Thirty or forty people came: Mary and the four children, my mother and father, and my brother, Chuck. Mary didn't say much; she let the relatives do the talking.

I didn't have any second thoughts. I was kind of flippant; I felt almost like a kid. Very easy about it, very light. Yes, tomorrow I would be going to the moon. A test pilot almost has to develop that kind of attitude. You don't want to dwell on the seriousness of the situation. Avoid possible complications. You have to stay on the light side. The kids were quite light about things too; they didn't take it very seriously that I was going to the moon. They had seen so many other guys go they figured I could do it too. Except Jan, the youngest girl; she had prayed every night that I wouldn't go to the moon. Well, it looked as if she was going to lose out.

I dozed off for a while. Just a little sack time. No dreams. I figured it was going to be a busy trip. Just as I got settled in good, we got the call. Deke was there monitoring it; the call came from the launch pad. Everything was on schedule. So we were disconnected from the console and we went one by one down the hall, each of us carrying a little portable ventilator in one hand. Being the shortest, I had to hold my hose high to keep it from dragging on the ground.

The elevator took us to ground level and we moved in single file out to the van that was waiting there. Newsmen, friends, relatives were waiting there too, with guards keeping them back. We lumbered along like elephants in the circus with the hoses dangling down.

It was early morning, about 6:30 or so. You couldn't hear any birds singing; all you could hear was the swish of the oxygen. We were already encapsuled, and it was interesting to see all those smiling faces that had come out to wish us well. We had a police escort, and we had a backup van following us, just in case the van we were in broke down for any reason. We drove down the highway, past the vertical assembly building, and on to launch control. When we stopped there, Deke Slayton made his final farewell, something like, "Have a good mission. Enjoy yourselves and have a good flight." As he looked us in the eye, we could tell that he'd like to be in our shoes. Now, only the suit technicians were with us as we continued out to the launch pad.

The van drove up the ramp by the launch pad and parked next to the elevator. There we were: Col. David R. Scott, Maj. Alfred M. Worden, and I. We had been extremely quiet. No clowning around, just very serious. Dave and Al were very cool, as always. We got out and into the elevator and started up. At 360 feet, the height of a thirty-six-story building, we got off the elevator and walked across the swing arm, which is sort of a catwalk. You can look down that gigantic missile, all the way down to the base, and you wonder how something that huge will ever get off the ground. Then we went into the white room—it surrounds the command module—and crew insertion began.

We were inserted in the spaceship, almost shoehorned in, one at a time. The suit techs helped us route the hoses so we were connected with the oxygen in the Command Module; they saw that we got onto the couches properly, and they cinched the straps, seat belts, and shoulder harnesses. Karl Henize, one of our support crewmen, was in there assisting with this operation and also looking as if he wished he were going along. Dave went in first; I was second, and Al third. The electrical connections were made so we could talk to Launch Control at the Cape and Mission Control in Houston.

There we were, on our backs with our legs up in what they call the 90-degree position. The suit technicians and Karl patted our shoulders and left the spacecraft. When they closed the hatch, it kinda clanged like a dungeon door. I think that is when the reality of the situation hit me: I realized I was cut off from the world. This was the moment I had been waiting for. It wouldn't be long now.

TO RULE THE NIGHT

We were poised on the very nose of a Saturn rocket made up of three main stages: the S-1, the S-2, and the S-4B. The rocket consists of rigid tanks full of fuel and oxidizer, a highly flammable combination. There was an abundance of light inside, reflecting off the light gray interior. The Command Module had as much space as a medium-sized bathroom. We were lying side by side like three men on a boardinghouse bed. There was additional room at the foot of the couches where a man could stand up and not interfere with the guys who were lying down. Actually, we all managed to stand up down there, in what we called the Lower Equipment Bay, or LEB, when we had to.

After they closed the hatch, there was not a great deal for me to do. We got in about 7 A.M. and launch was about 9, and we were kinda tired because we got up early, so I dozed off a few times. I was reflecting on my life, wondering what was in store for me, wondering whether I had made the right decision. My life flashed before my eyes. When I'm asked, I say, "No, I wasn't frightened. Astronauts are never frightened," but I hasten to add that during those hours on the launch pad there was an air of anticipation and expectancy as we waited for ignition. And then—with my feet up in the stirrups and my back on the pad—I had cause to reflect on a little problem I had had in training, urinating in this position. The first time the connection must have come loose and I had been drenched. When I got back to the training building that day I had lain on the floor and practiced, trying to get the knack of it. Well, just as I had feared, as I lay there that morning I could feel the need for relief. Thanks to my earlier training, I overcame the problem. I was ready for launch.

The time had been dragging, but the last minutes went

very fast. Before we knew it, we heard the word "ignition." We sensed and then heard all that tremendous power being released underneath us on the pad. Slowly, tremulously, the rocket began to stir.

We knew that if we cleared the tower we had a reasonable chance of survival if something should go wrong. I watched all the systems I was responsible for on my side of the spacecraft. We cleared the tower. It was almost the happiest moment of my life to realize that after all those years it was now my turn. At last I was leaving the earth.

The muffled roar flows through you. You just hang there. Then you sense a little motion, a little vibration, and you start to move. Once you realize you are moving, there is a complete release of tensions. Slowly, slowly, then faster and faster; you feel all that power underneath you.

We are building to 4 G's, riding the center of this roaring force until we pass through Mach 1; then the noise will drop off. I am concentrating on my systems, lying on my back, with my console almost overhead in front of me. I'm taking the most thrilling ride you can imagine. A lateral oscillation gives you the sensation of taking a curve on a high-speed express train; it is a combination of up and continually pitching over before you finally get horizontal. As you build up to 4 G's, you weigh four times as much as you do on the earth, and you are plastered against the couch. You are pressed back against the couch under fantastic weight, and it is difficult to raise your arms to touch a switch or move a lever. It would be only in an abort situation that you would have to do this—and my chore, primarily, during the launch is to stand by for such an emergency. If everything is normal, I just go along for the ride, checking the instruments to be sure we are on the right trajectory.

Just then you come into staging and the engine shuts down—WHAM! All of a sudden you are thrown forward against your straps. It feels as if you are going to go right into the instrument panel; you unconsciously put your hands up to absorb the impact. You are holding, just lying there. The engine shuts down, the structure unloads, and the spent stage drops off. That's a hundred feet of rocket dropping off. After an interval of a few seconds the next stage lights off —BAM! You are pushed back on the couch again. You have built your 4 G's and you are going faster and you stay at 4 G's. It's like an afterburner on a fighter; you just keep going faster and faster and faster. Then there is an interval of a few seconds when you just hang there. The guys who briefed us told us that when you go through staging it feels like a train wreck. Now I know what a train wreck must feel like.

For the first staging you are visible from the ground; for the second you are not. You maintain this incredible high-speed ride, gradually pitching over as you accelerate. By the time you reach orbital speed of about 18,000 miles per hour, the spacecraft has been horizontal for about two minutes.

Of course we could see out the windows because the protective cover was blown off the Command Module at about 15,000 feet, after the danger of an abort was past. (Until that point there is a possibility that you might have to use the launch escape tower to pull you off, and the exhaust might damage the windows if they were exposed.) After the cover blew we were going straight up, and we saw blue sky that got blacker and blacker. Once into earth orbit we could see blue skies below and black skies above. Right in the middle of my window was a full moon. Seeing the full moon was a terrific omen. I knew we were going to have a great mission. At the moment of insertion into orbit, when the engines had shut down, our destination was framed in the window and the whole flight had this absolute clarity.

From this point on, we were very busy. We had to check out all the systems in the Command Module to be sure everything was working the way it should so that the Trans-Lunar Injection (TLI) Burn over the Pacific would put us on a trajectory to the moon. This involved igniting our rocket engine and firing it for approximately six minutes. We couldn't take time to look at the continents below us or the beautiful sunsets and sunrises as we orbited the earth every hour and a half. We were moving from light to dark to light with a tremendous sensation of speed. Our days were forty-five minutes long, our speed about 18,000 miles per hour. This was our first real experience with weightlessness.

I had to accept the fact that I didn't feel completely comfortable. It was amazing how Al was scooting around—he had to go down to the Lower Equipment Bay to make his navigational sightings with the telescope and sextant. There was no need for Dave or me to move because we could do all our tasks from our couches. We were weightless, so we loosened the straps and used them just to maintain the right position. I didn't want to do any somersaults, or even try to move my head fast. I felt that I might have motion sickness if I made any sudden moves. We had taken our helmets off by this time, so if we became nauseated it wouldn't have been serious, but it would have been miserable. If we threw up we'd have the problem of containing the stuff.

We were weightless, the pressure was 5 psi, and the atmosphere was essentially 100 percent oxygen. There was a little nitrogen that we were trying to purge. We were all busy as we approached the time for the TLI Burn. As we passed over the tracking stations on the earth, we got GO signals. Just as we got the final GO for the TLI, I looked out my side window and there were the Hawaiian Islands, all of them in the blue Pacific Ocean, framed right in my window. I could even identify the high volcano mountains of Mauna Loa and Mauna Kea. I wished I had my camera so I could get this picture. We were a hundred miles up looking down on mountains I had been over on foot.

We were ready for the burn. We lit the engine, and it seemed as if we were going straight up. Again, what tremendous exhilaration to be lifted up, to leave the earth behind. During what seemed to be just a few minutes, we did a couple of things in the spacecraft and then we looked out the same window and there was the entire southeastern part of the United States spread out before us—the peninsula of Florida, just a few inches long, and Cape Kennedy. There were few clouds, so I could see Cuba, the Bahama Islands, and those beautiful blue-green waters off the coast of Florida.

I did have one earth thought: I wondered if they were still fighting the traffic jam to get back from the launch site to the motels. We had been out about two and a half hours, we had gone around the earth a few times, and I wondered if they were still trying to drive from the cape to the motels.

We were all so busy in the spacecraft that it must have been three or four hours before we had a chance to maneuver the craft and look at the earth again. When we did, we could see the full earth—North and South America, Europe, Africa. You could see the blues and greens, the tans of the deserts, and the whites of a few clouds, and there was black all around the earth. You couldn't see any band of atmosphere, no blue at all. It was the full earth with the sun shining right on it, fully illuminated against the blackness of space.

We couldn't see the earth as an oblate spheroid; it looked

perfectly round. Then we had to maneuver the spacecraft again and take pictures of the moon. Since the burn over Hawaii, we had been on our trajectory to the moon. We were up to 25,000 miles per hour, but the spacecraft would gradually decelerate until we reached the influence of the moon. We did not have the feeling of speed.

While we were in contact with earth we knew that the earth could monitor the systems of our spacecraft better than we could. If a gauge indicated a trend that suggested a problem, Houston would usually pick it up before we did. All manner of redundancy is built into the spacecraft, so we could select an alternative system to get around a problem or we could try to repair the system. We could take panels off, and we had a tool kit for simple repairs, but if an electrical problem developed there was not a lot that we could do. Most of the electrical packages were self-contained boxes. You wouldn't want to dig into any of these boxes to try to replace any parts. We didn't have any replacement parts in the spacecraft, anyway.

If one system failed we had to depend on doing the job another way. Of course, when we got to the point where the Lunar Module was sitting on the surface of the moon, there was just one engine. If it didn't fire, that would be it. There is no way to land the Command Module on the moon and rescue the crew from the Lunar Module; the Command Module doesn't have landing capability, and the Lunar Module will not fly you home. You can't get back into the earth's atmosphere without the shielding of the Command Module; to withstand temperatures of up to 5,000 degrees F, the Command Module was covered with several inches of ablative material that would melt off slowly under the intense heat.

There were some hairy moments during the flight, but

with the help of Houston we came through them in good order. However, there were some housekeeping problems that we weren't prepared for that were pretty aggravating. Living in those cramped quarters in space is pretty much like being a mother in a small apartment with three sick children on a rainy day.

Just taking your space suits off—it's like three people dressing for a formal dinner in a broom closet. And these suits cost more than $60,000 apiece. You have to help each other, or you run the risk of damaging the spacecraft. Flailing around in there you could hit a gauge or a switch or a circuit breaker. There is not so much risk of damaging the suit, but with all the zipping and unzipping, it's a lot safer if you zip and unzip each other and pull each other's suits off. Then you have to do something with the suits—empty, they are almost the size of a man. It's almost like having six people in there instead of three.

That first day we were in our suits six or seven hours, and it was a welcome change to take the bulky things off. Going about the most humdrum chores is weird when you are weightless. The sensation is hard to describe, but if you have ever done scuba diving, or any diving where you have been ballasted so that you are free to float at a particular depth, you have experienced a reasonable simulation of weightlessness. When there is no up or down, you become disoriented, and there is a tendency for your stomach to become disoriented too. Anyway, think of doing somersaults in this state. It actually took me about three days to get used to it, and by then it was time to land on the moon and get used to 1/6 G. That is only one sixth of the gravity of earth, but it is a tremendous comfort if you haven't had any gravity at all for four days.

If you want to move from one place to another in a

spaceship, you just push off with your fingers and float right across. The first day or so, we'd always overdo it. We'd push too hard and careen from one side to the other, colliding with each other. And, besides, everything that gets loose floats too. Before you know it everything is in the holding pattern together.

Think about the simpler routines of housekeeping. We each had two containers for our personal belongings. We had a long bag with a snap top strapped to the bulkhead by our couch, and we had another container up in the tunnel area where we kept our toilet articles. There were thousands of items in the spacecraft that had to do with the scientific experiments programmed for the flight. We had parts and accessories for a Nikon, a couple of Hasselblads, and a movie camera. We had sextants, telescopes, other navigational equipment; there were tools, devices, equipage of every kind, and instructions for everything. We soon discovered that the most important items aboard were adhesive tape and scissors. And then we had food.

When it was announced that the crew was eating, everybody thought that we must be relaxing. Believe me, it was not that way at all. First you've got to find the meal for that particular time of that particular day. Of course we had a stowage map which told us where everything was when we left earth. All food containers were labeled A, B, C, D, E, and all meals were color-coded. Mine were blue, Dave's red, Al's white.

So if a meal floated by and you identified it as being red, you could say, "Hey, Dave, you've lost your entree." Actually, we had to improvise a way of managing this flying circus—it worked out best if I prepared the food. Three people can't use one water gun at the same time to mix different amounts of hot or cold water with ingredients of eighteen bags. I read the directions—six ounces, eight ounces, hot or cold. And then after I shot the water to the bag, I'd wait five or ten minutes for the bulk to absorb the water. Or, I'd float it out and let the customer age his own bag. Any way you did it, this had to be the most unusual small restaurant in the world or out of it.

We had Velcro all over the spacecraft and on the meal packages, so to keep track of things we'd stick our dinner on the wall, course by course. If you nudged the meat course accidentally, it would take off, and you would have to float after it or get the help of a buddy downfield.

One meal for one man comes in a package about the size of a large box of cornflakes. Inside that package are about six plastic bags with dehydrated food which has to be rehydrated. But first you've got to break into your meal. So you take your private pair of scissors and cut through the outer package, which is Teflon-coated, to get to the inner bags. Then you unwrap these inner bags, one at a time, and locate the spout at the end of the bag. You cut it off, exposing a valve, and you insert the nozzle of the water gun into the valve and inject the proper amount of water, per instructions. When your bag is ready, you cut into the other end and unravel it to get at the nozzle and pop it into your mouth. Then you roll the bag like a toothpaste tube and squeeze out the contents.

The foods that needed water for reconstitution had a fluid consistency like gruel. However, our soups were packaged in a different way, and eating them required some acrobatics. They were also in plastic bags, but they had a Teflon seal that you had to peel off. We added water to the soups, then very carefully pulled the tab to open them up. If you opened them slowly, invariably the soup would start coming out in bubbles or blobs that would float all over the place. The trick was to open the bag fast so that the viscosity or capillary action would encourage the soup to adhere to the plastic. The object was to take advantage of whatever adhesiveness the soup had.

While the soup was tentatively contained, you could actually eat it with a spoon under weightless conditions. You would put the spoon in there and sort of direct the soup toward your mouth, hoping that it would stay on the spoon. If your aim was bad, the soup would come off and float by. If it came off, you might as well get a towel and run the blob down and soak it up quickly. When you are eating you want to relax; you don't want to be chasing blobs of soup.

The gourmet soups were worth the effort; they were delicious. I had this favorite place down at Lake Wales, Florida, called Chalet Suzanne. I had gone over there many times during astronaut training, and Carl and Vita Hinshaw, who run the restaurant, were good friends of mine. They had wonderful food, and I enjoyed the atmosphere. Well, the Hinshaws were anxious to get one of their products on the flight. So, I convinced NASA that we ought to have these soups on board, and Chalet Suzanne met the rigid qualifications that NASA required. They freeze-dried and packed lobster bisque, crab, mushroom, vichyssoise, gazpacho, and romaine—imagine it: three men in long underwear flying to the moon and sucking in those blobs of lobster bisque.

The soups were probably the most popular food that we had on board. We also had a variety of meats in aluminum foil—ham, frankfurters, turkey, and steak, in flat slices except for the frankfurters. The meats had different consistencies, but all had the same taste. When we cut open the end of the aluminum package and slid the meat out, the gravy or grease would slide out in blobs and float around the spacecraft. We had real air pollution in there—and no air outside at all.

Dave was very pollution-conscious, and he kept reminding everybody that a clean ship was a happy ship. When we had finished eating, we had food all over us like a bunch of two-year-olds. So we'd take our color-coded towels, give them a shot of hot water, and begin mopping ourselves off. We were still wearing only the long underwear, because it was warm in the spacecraft, and soon the upper portions began to look like well-used bibs. Our fancy Beta-cloth overgarments—high-style white coveralls with feet like children's pajamas—were stashed away. They would have been great if we had been required to make a social appearance during the trip, but, things being the way they were, we never used them.

The food packages were very compressed when we started out, but after you finish eating the food you end up with packages that are twice as big, and you have the problem of stowing them away.

This involved cutting open another bag, breaking out our supply of yellow bacteria pills, and putting a few into each bag along with remnants of food. Then we would wind up each bag like a spent tube of toothpaste and shove them into the large plastic bag and pop that into a container. You had to squeeze everything real tight, or the bags would fill up with oxygen and billow out. We didn't have room for that, and we didn't have room to grow a crop of bacteria either. Sometimes, with all the bags around in different stages of the cycle, it looked as if the Monday wash were hanging out. Housekeeping! Nothing but housekeeping!

Soon we began to notice another effect of weightlessness. Our heads were swelling. We could feel our faces getting larger, and we could see each other turning red. It was a new condition for the body, and the brain was accommodating itself to a different sort of blood pressure and to pure oxygen. It took a couple of days before our bodies adjusted and the swelling started to recede.

Curiously enough, the brain seems to work better in space. We all found that we worked more efficiently in space than we did on the ground. There was great clarity of mind and, as I have said before, something akin to clairvoyance. The doctors haven't suggested this, but it may be partly because the brain is getting more oxygen.

We didn't get to bed until 10 or 11 that night, which made it a long day, since we had gotten up at 4:30. Sleeping in the spacecraft was another challenge that required special skills, but before this there were chores that had to be done.

We had to add chlorine to the water in the Command Module every night to keep the growth of bacteria down. We also had to change the lithium hydroxide cartridges which kept our oxygen purified. Then came bedtime.

We knew that getting enough sleep was a serious matter, and we had practiced sleeping with a recorder playing the vibration noises of a spaceship, but we had to do a little improvising when it came to actually bedding down.

Getting ready for bed involved some practical problems. When you dump urine or water, it immediately forms ice crystals. And since these ice crystals are traveling at the same velocity as the spacecraft, you move in a cloud of beautiful crystals—each with a different shape, and in every color of the rainbow. Your humble waste product has become a radiance, but this presence interferes with the observations of the stars that we needed for navigation.

41

Inside the spacecraft there is a more immediate consideration: blobs of liquid might get inside the instrument panels and damage the delicate circuitry.

I don't know whether it was because of high fluid intake or weightlessness, but I had to urinate every hour. We did consume a lot of liquid, since our food was reconstituted with water and we had plenty of fruit drinks. During most of the hours in flight, we voided into a yellow collection bag. You fitted yourself into a hose attached to the bag, then opened the valve and hoped for the best.

The basic sleeping arrangement was the three couches in a row. The Command Module's consoles for monitoring and operating the craft were overhead and reachable when we were strapped down or tethered to the couches. This was fine for burns, takeoffs, and landing. But as for sleeping there grown men in a row, it was like three grown men in a double bed. Every time one man rolled over he would disturb the man next to him. So we did a little improvising. Al used the left couch, Dave used the right one, and I slept in the space underneath Dave's couch. My little cave had an area of about two feet by six feet, and it handily contained a combination hammock and sleeping bag.

I could slide the lower part of my body into this sheath, but it wasn't quite long enough; the upper part of my body would hang out. I spent the first night floating or drifting around in my little space, sleeping fitfully. I would wake up with my head cocked at a weird angle and then suddenly realize where I was. It was very disturbing.

I knew I had to do something about this sensation of being unanchored, so I wadded up one of the trash containers and made a pillow out of it; then I stuffed it between two struts. In that way I wedged in my head, so it wouldn't drift all over the place. I remember hanging onto the pillow, feeling almost like I was on the earth.

Before we went to sleep we had to put metal window shades over the windows to block out the light. We kept the spacecraft in a rolling motion, or barbecue mode, to distribute the heat evenly around the vehicle, and when the windows would rotate into the sun, an intense light would come in that was actually blinding. Then, as the craft turned, the windows would be looking out into the blackness of space. In space you have the extremes of temperature, since the part of an object that is in the sun is 250 degrees F and the side away from the sun is minus 250 degrees F. With the shades up, it was black in there, and we were not aware of the roll.

As we got accustomed to weightlessness and to the peculiar noises of the spacecraft, it was very good sleeping in space. Houston woke us up every morning—and they could tell what our state of wakefulness was because they were monitoring our respiration and heartbeat. We took turns sleeping with the headset, so that we would always be in contact.

We had a lot to do, and some chores could not be done very rapidly in space. Going to the bathroom is a prime example. Even though I went through a low-residue diet for three days before flight, it didn't seem to make the slightest bit of difference. My earth habits persisted in space. The first morning I woke up I knew I had to go and I dreaded it. Defecation in space is an art I could never entirely master. We had practiced with plastic bags at Dave's suggestion, but of course we had never done it at zero G's. It meant taking off all your clothes and going down to the Lower Equipment Bay with this plastic bag that has an especially designed top

43

with a round receiver and a flat rim. You peel off a circle of tape with a sticky surface that you put right on your bottom.

The whole hygiene business was really aggravating. I began to smell like a restroom, and it really got to me. I've always been sort of a shower nut, a compulsively clean person. To be in an environment where I couldn't clean up and had only three sets of long underwear—it got so I couldn't stand my own company. I got scroungier and scroungier as the trip went on.

We brushed our teeth after each meal. I had taken a razor because my face starts itching when I grow a beard. But Dave and Al hadn't even brought razors, so I decided that if they weren't going to shave, neither would I.

Bathing was another chore. We would dampen a washcloth and clean ourselves all over as well as we could. For some reason, the other guys didn't bring any soap. Fortunately I took a bar of sweet-smelling soap along. It was the high point of the day just to take out the soap from a container and let the scent waft around the spacecraft. It almost made us feel clean.

The second morning out we did some chores before breakfast. Ever since shortly after the Trans-Lunar Injection Burn we had been aware of a short in the Service Propulsion System (SPS) engine. We caught this when we noticed a warning light.

The SPS engine is the rocket engine for the service module. You use it for all your burns: mid-course burns, the burn into lunar orbit, and the burn out of lunar orbit. It's your ticket home. Any doubt as to whether we could fire this engine raised the question as to whether we still had a GO for continuing the trip. At 10:09 A.M. Houston had some definite conclusions about the location of the short: CapCom (Capsule Communicator) at Houston, our voice contact in

Mission Control who managed all communication with us, said, "We're interested in finding out where in your Delta V thrust A switch the short seems to be, and all of us down here are convinced that it's either in that switch or physically very near that switch."

One of the four scheduled mid-course burns was set for 3:26 p.m. that afternoon anyway. If the early test proved that the engine worked, there would be nothing to worry about. With this exercise in prospect, we ate our first meal of the day at 10:30 a.m. We didn't exactly have a seated affair, but we did have breakfast.

When the time came for the test, Houston instructed us to tap on the panel where we had discovered the mysterious light. We had to have the light back on, i.e., the short active, to test the adequacy of backup circuitry to do the job. We got the light. At the decisive moment, the computer gave the engine the signal to start. It fired off perfectly, and after a burn of less than a second the SPS engine was shut down.

The test of the Service Propulsion System worked so well that the scheduled mid-course correction burn number 2 was scrubbed. The trajectory was perfect; the little test had effected the correction. Houston reported that at the time of the burn we were 114,787 nautical miles out from earth.

The second day after launch continued lively. That evening we made a scheduled inspection of the Lunar Module a little before 6:30, to see if it had been damaged in any way during the flight. When we got in, we found there had indeed been some damage; there were particles of glass floating around all over the place. We found that the face of the tape meter had been shattered. The fragmented portion, which was about two inches wide and six inches long, had broken down into a number of slivers. Free-floating glass is the worst thing to deal with; it can get into your eyes or into your lungs. There was a real danger that it might get into the Command Module, where tiny particles could get behind the panels through cracks and loose fittings and damage the instruments.

It could be a problem if there was any damage to the tape meter itself, because this was a critical instrument for landing the Lunar Module. During the rest of the flight we were involved in a continuous chase after glass particles.

We hurriedly shut the hatch between the Command Module and the Lunar Module and went after the big particles with gray adhesive tape. Like the scissors, it was a secret weapon. We had rolls of the tape; we'd strip it off to make a kind of sticky fielder's glove, and we'd use it to mend or reinforce things that were damaged or broken. In this new, curious, weightless world, it was a surface that you could count on to hold things.

It is amazing how restless everything was. If our snap-top bags were open the slightest bit, things would start drifting out. Toothbrushes, Kleenex, cameras, all sorts of things were floating around. Some of the screws that fastened my seat belt worked loose, and occasionally we'd see a screw float by. The tremendous accelerations and, I think, the constant low-level vibration worked things loose.

We had a cabin fan over the left couch that had a screened inlet, and all the oxygen in the spaceship was sucked through this intake and recirculated. Eventually everything would end up there. At the end of the day, we would run our trap line: go over to the inlet with a piece of our gray tape and collect all our oddments. We would identify the ones that were valuable and put them back where they came from. The stuff that was useless we put in the trash.

That night after we had gone to bed as usual, Apollo Control in Houston briefed the nation and tucked us in as far as every soul listening on earth was concerned. Not only did we feel plugged in to Houston, we felt plugged in to all mankind. It was a strange and wonderful feeling.

The next morning Houston awakened us at about 9:35, an hour later than the planned reveille. Everything was evidently okay. When the "Good morning Apollo 15" greeting came from CapCom, Al responded in high spirits, "Okay, Joe, we certainly did have a nice sleep and we think your tracking data must be right; the moon is getting bigger out the window." I was a little late rallying that morning. When CapCom said he was ready to read us the *Gold Bugle Morning News*, Al Worden stopped him. "Stand by for a minute on that," he said. "We wouldn't want our LMP [Lunar Module Pilot, that's me] to miss it." So they got me up, and I started a brand-new day in flight.

Houston was right there with instruction on matters domestic, as well as scientific. It is incredible how much housekeeping there is at the outer reaches of technology. CapCom listed the chores: ". . . Vacuum the cabin and the filter, and the subgroup under this. Unsnap the netting around the cabin fan filter, then you vacuum the filter, but do *not* scrub the bristles of the vacuum cleaner over the surface of the filter. Then," CapCom went on, "you remove the remaining particles on the cabin fan filter with the sticky paste [meaning tape], using care not to dislodge the filter material. Then you remove the particles on the inner screen of the vacuum cleaner with sticky tape and, finally, replace the netting."

Then came the Visual Light Flash Phenomenon Experiment, which required that we cover the windows of the spacecraft and put on eyeshades. One hypothesis was that

47

the flashes seen by earlier Apollo crews were visual phosphenes caused by cosmic rays. There was some question as to whether the flashes were caused by high-energy particles traveling through the eyeballs or colliding in the retina of the eye or cerebral cortex, the visual center of the brain. Lying there in the dark, flat on our backs with those little blinders on was so relaxing that I must confess I dozed off a few times.

In some instances the cosmic rays are brilliant. They are as definite as the flare of a flashbulb on the other side of a dark arena. In some cases, all three of us would see the same flash. Evidently the same cosmic ray would go through our three heads.

The most relevant concern for space travelers was the possibility of lasting effects from cosmic rays. Would this bombardment destroy the vision of a man who spends, say, a year or a year and a half in space? Would it destroy his brain? Houston took pictures of our eyeballs before and after the experiments, and they couldn't see any change in us. They are still watching. Who knows?

That afternoon Dave Scott and I went into the Lunar Module a little after 5:30 P.M. and worked there for about two hours. Everything seemed to be fine in the LM, and all of the systems checked out good for Houston. We reported later that we had collected an estimated 60 to 70 percent of the broken glass. After tests on the tape meter to see how it would operate in pure oxygen with its broken front, Houston reported that they anticipated no problem. Everything seemed to be shaping up for a successful arrival on the moon.

In the evening I went down into the Lower Equipment Bay to chlorinate the water again. I was taking the chlorination injection kit off the inlet fitting when the water started to come out, a little at a time. I told Dave it looked like we had a leak. He went down immediately and examined it. No question, the leak was getting worse. We felt at the time that maybe the line had cracked; if that had happened, there would be almost no way of fixing it. So we told Houston about the problem. Houston asked, "How many drips per minute?"

I wish that we had such things as drips in space, but without any up or down, water doesn't drop out. It just makes a blob, and the blob keeps getting larger and larger. We had to get this water contained somehow, so we started looking for towels to mop it up. And wouldn't you know, about that time the locker with all the towels in it jammed. Al and I were trying to get the locker unjammed, Dave was trying to contain the water, and CapCom was still trying to get an estimate of the volume we were losing.

"Yeah, it's a pretty good flow right now," Dave told them. "Drips per second is hard to measure, but there is a whole ball of water right around the valve."

Karl Henize, the Capsule Communicator at the time, said, "Dave, I had a problem when I chlorinated on launch day, and when I first took the valve off I had about what you've got—quite a strong flow. The cap stops it from flowing when you put it back on, and after I chlorinated, the flow decreased down to a very slow drip, say once a minute."

"This is a big run, Karl," Dave said intently, "and the cap is on tight and you can almost feel something flowing beneath the cap."

"Okay, stand by. Lots of people are thinking down here now," Karl said.

After a few minutes CapCom made a diagnosis:

HOUSTON: We suspect that the injector outlet is loose and we have a procedure here for tightening it up.
APOLLO 15: Okay, give it quick.

49

HOUSTON: Roger, we need tool number three and tool number W out of the tool kit.

APOLLO: Okay, three and W out of the tool kit.

HOUSTON: Right, put number three in the tool W ratchet and insert tool three in the hex opening in the chlorine injector port.

APOLLO: Okay, that looks like where it's probably leaking.

Houston's final suggestion: "Once you've got tool number three well engaged in that injection port, turn it about a quarter of a turn."

It stopped the leak.

What really concerned us was the thought that if the fitting had broken, or if we had been unable to stop the leak, we could never have landed on the moon. If water were floating around, it could get behind any of the panels and short out the electrical devices.

Well, we got the jammed locker open, and took out all the towels we could find. With towels, our dirty clothes, and everything else at hand, we mopped up a quart-sized blob of water. When we finished we had wet clothes strung out over the Lower Equipment Bay and up in the tunnel area going into the Lunar Module. It looked like an old-fashioned clothesline.

We are in training for years to meet any eventuality, and then we are hit with a plumbing crisis. When we go into space we always know there will be some problem that we haven't anticipated, but we never know what it will be. I guess it's Murphy's law. If anything can go wrong it will. But good old Houston came through again.

Throughout the trip we benefited from Dave's thorough knowledge of the Command Module. He had flown twice before, on Gemini 8 and Apollo 9. The Lunar Module was my specialty.

NASA realized early in the program that it was too much for one guy to learn both vehicles completely. They gave us the opportunity to make a choice, and then we specialized. Early in the program, I decided to take the Lunar Module. This would give me an improved chance of going to the moon, and it's the more exciting vehicle because you do more flying.

My background probably had something to do with it. The Lunar Module is more like a fighter aircraft, while the Command Module is more like a bomber in its responses. There was a time, if somebody stopped me on the street and said, "Who are you?" I might say "Jim Irwin" or I might just as likely say "The Apollo 15 Lunar Module Pilot."

That third night after we had gone to sleep, the spacecraft passed into the lunar sphere of influence, and the moon became the point of reference in computing our position.

We are sleeping soundly when Houston calls to wake us up a little after 6:30 Central Daylight or Houston time. "Good morning, Dave. It's time to rise and shine." There is expectancy in the voice at Houston and excitement out in space—the moon is ahead.

We cannot see any light on the moon's surface. You look out and there is a dark object looming up, a big mass in the darkness of night. You have no concept at all of its features because of this blackness. The sun is on the other side. At this point we are not in position to see the earth. People think you can always look out and see the moon here and the earth there. But you can't unless the windows are positioned just right.

51

At about 12:41 we see a very thin crescent moon in front of us. Despite the delicacy of the shape, we get the impression that the moon is very big. We are silent as we coast in to the moon. We are preparing for the Lunar Orbit Insertion (LOI) Burn to slow us down so that we don't leave the moon behind. Our course is taking us behind the moon, and when this happens we will lose contact with the earth.

"Have a good burn," Houston says. And we say, "We'll see you on the other side." For the first time we are behind the moon and out of touch; we are on our own. This is the sort of privacy I like. We have a great burn. We fire the SPS engine that slows us down to about 3,500 miles per hour. When we come out of it we are coasting in lunar orbit.

All of a sudden we come from darkness into daylight. You are at the moon! It hits you just like that. It is the most beautiful sight to look out and see this tremendously large planet. You'd never guess that the moon would be that big, even though you have seen all the pictures. But here you are seeing it with your own eyes for the first time. It is staggering. You can barely see the curvature of it. You are coming around and moving not too fast, at medium speed, and you cross the terminator, which is that line between darkness and light.

The surface of the moon is a dark gray, gunmetal gray. It looks like molten lead that has been shot with BBs. It doesn't look real; it looks like clay. First it goes from black to dark gray. Since there is virtually no atmosphere on the moon to diffuse the light, there is a very distinct line between darkness and light. As the reflected sunlight gets brighter, the gray turns to brown, light tan, and almost white, directly underneath the sun. So you have this constantly changing beauty and color as you go around the moon. The moon is turning, and our orbit is changing as we go around. As we shift from north to south, the scenery is endlessly different.

Uncharacteristically, Dave can't contain himself.

"And you know as we look at all this after the many months we've been studying the moon, and learning all the technical features and names and everything, why, when you get it all at once—it's absolutely overwhelming. There are so many different things down there, and such a great variety of land forms and stratigraphy and albedo that it's hard for the mental computer to sort it all out and give it back to you. I hope over the next few days we can get our minds organized and be a little more precise on what we're seeing. But I'll tell you this is absolutely mind-boggling up here."

Old slow-walking, slow-talking Jim didn't say much, but I was knocked out by what I was seeing. It was amazing. Coming from the back side, the first large feature that we saw was Tsiolkovsky Crater, with its high central peak of light-colored material surrounded by a dark sea. The crater must have been fifty miles in diameter and probably as deep as the Grand Canyon.

The back side of the moon is distinctly different from the front; it has no basins, none of the flat surfaces that you see on the front of the moon with basins, like Mare Serenitatis and Mare Imbrium. Flying over the front you see craters of every size and description; you see mountain ranges, canyons, or sinuous rilles—you could gaze at it hour after hour, particularly when you get to the terminator. There you pick up the sharp relief because of the long shadows; it makes the mountains stand up higher. The line between darkness and light sort of jags around, depending on the topography.

It was a little confusing to me at first because it seemed as if we were flying over upside down and backward. This is disorienting. I am surprised that Dave and Al were able to pick out the land features as well as they did. I don't know why we were in that particular attitude, unless it was to point the scientific-equipment bay at the surface. We call the point on the moon directly beneath the sun the subsolar point, and as we moved away from this point and it began getting darker and darker, we flew over the Apennine Mountains. We could look down, and there was Hadley Rille. This is the spot where we were going to land. We could see the fantastic canyon, but the landing site at that time was actually in the shadow of Mount Hadley, highest mountain in that basin, towering 15,000 feet above the surface. Mount Hadley is twice as high as Pike's Peak from base to peak.

The whole panorama spun below us every two hours as we orbited the moon. We were looking down at some strange territory when it was what we call moon-noon. With the sun directly overhead, it was 250 degrees F. I don't think we could survive that, even in the Lunar Module. But we didn't have to, because we were scheduled to land in the early morning and leave before noon. Although we planned to spend three days on the surface of the moon, this was easy because these were "earth" days. One moon day is equal to twenty-eight earth days. So we could land in early light, spend three days, and get off before 9 A.M. in moon terms.

We had to wait until the shadows receded from our landing site, so we could see what we were doing. Then we would fly in over the mountains and land when the sun was 12 degrees above the horizon. The temperature would be about 130 degrees; before we left, it would rise to 180 degrees. We were beginning to get the impression of speed as we moved across the lunar features of the moon. Houston was aware of this; it seemed almost as if they were in the spacecraft with us.

CapCom says, "Fifteen, does it look like you are going to clear the mountain range ahead?"

"Karl, we've all got our eyes closed; we're pulling our feet up," Dave says.

"Open your eyes. That's like going to the Grand Canyon and not looking."

It was a fantastic overflight. I was fascinated by the majesty of the lurain. Here I was right over the mountains of the moon—I could see their beauty and I felt I could almost reach down and touch them. I was exhilarated when I went to bed that night.

Houston woke us up at about 6:15. We got the call, and we got moving. By 10 we began suiting up. Since there is always a risk when you undock that you might lose pressurization in one of the vehicles, all three of us put our space suits on. Then we could make a vacuum transfer if the need should arise. Dave and I went over into the Lunar Module, where we had more room to work, and each of us zipped the other. Then we fired up the Lunar Module and closed the hatch. We were all standing by for the undocking. Everything was going smoothly, but then came the hitch: the Lunar Module wouldn't undock.

Dave and I looked at each other. We couldn't figure it out. As we were puzzling over it, Dave told Al to go back into the tunnel to check the connections. It could have been an umbilical that was not properly connected, in which case power would not flow to the probe which had to be actuated to effect the disengagement of the Lunar Module. When we came around the corner of the moon, Houston gave us about forty minutes' grace before we would have problems in making the descent as scheduled. Al meanwhile had to go through the tiring and time-consuming procedure of opening the hatch again, so that he could check the connections.

Al evidently had not had the connections firmly mated. He pushed the connectors in solidly, tried it again, and we undocked. Of course we were late undocking, but as it turned out it worked perfectly. We made it just shortly before we came over the landing site. Al fired the thrusters in the Command Module to increase the separation rate to about one foot a second. He maneuvered to get out of the way and into position for a burn that would put the *Endeavour* into a more or less circular orbit. And Dave and I floated right over the landing site. Now the *Endeavour* appeared to be ahead of and below our Lunar Module, *Falcon*. I took movies of us coming up on the Apennine Mountains, right over our landing site. Then we moved into darkness.

Now we are down in a 9- by 45-mile orbit—that is, 9 miles above the average surface of the moon at our lowest point. There are many mountains higher than that. You look out on the horizon and you see these high peaks and you are just skimming along. Now you really know you are moving fast. You are traveling about 5,000 feet per second, that's Mach 5 or 3,000 miles per hour. Your orbit is defined; you can't dodge anything. You don't have any control over the vehicle, and if you did you probably couldn't react fast enough. You just assume that Houston knows where the mountains are and how high they are. But you see the high mountains on the horizon and you move toward them very fast. You wonder if you are going to clear them.

The face of the moon is beautiful in a stark, awesome, barren way. It is all ochers, tans, golds, whites, grays, browns —no greens, no blues. We were hanging loose, coming to the burn itself. We were going down to land on the surface. Dave was doing most of the hand-control action, but in the main we were telling the computer what to do.

The Lunar Module is almost as big inside as the Command Module. You need room for two guys to stand side by side and look out the window. The area to the rear of the crew is not very wide, perhaps two feet, but long enough for a man to sleep in. Together we were sharing a space a little smaller than a bedroom on a train. The consoles were forward of us, right next to the window. We also had a console at our waist level, and we had switches and circuit breakers on the outboard side of both of us. We had redundancy in controls so that either of us could fly it. Actually, the Lunar Module flies very much like a helicopter; that's why we trained in choppers.

It was time for us to go into our Power Descent Initiation Burn, a twelve-minute power burn to the lunar surface. At the onset of this burn, the *Endeavour* would be about 350 nautical miles behind us; it would catch up and be approximately overhead at the time that we touched down. We started firing the engine, which has approximately the same thrust as the SPS engine of 30,000 pounds.

At 8,000 feet we pitched over, changing our altitude about 30 degrees. We had been coming in on our backs, looking up, feet first. Now we got our first good view of the landing site. After we pitched over it *was* like a helicopter. Dave looked to the left and saw a mountain up above us—it sort of misled him. In our simulations we had never seen a mountain above us. There we were at 7,000 or 8,000 feet, and Mount Hadley Delta was towering above us, soaring up 13,000 feet from its base. It gave the impression that we were a little short of our landing point.

Dave didn't want me looking at the surface at all; he

wanted me to concentrate on the information on the computer and other instruments. He wanted to be certain that he had instant information relayed to him. He was going to pick out the landmarks. But Dave couldn't identify the landmarks; the features on the real surface didn't look like the ones we had trained with. We could see the great canyon, or Hadley Rille—so he played the landing point to be about the right relative distance from it. There were lots of craters out there, making it difficult to find a smooth place to land.

I kept telling myself, "Jim, this is really a simulation. You are not really landing on the moon." If I had believed I was landing on the moon, I would have been so excited I don't know if I could have made it. It was really hard not to look out. It was the smoothest "simulation" we had ever run. Everything worked just perfectly. When we pitched over, the computer knew exactly where it was going to land the vehicle. It displayed a number in a keyboard. I gave that number to Dave, and he looked at a grid imprinted on his window. He located that number on the grid, and he could sight the landing spot on the face of the moon. The spot didn't look familiar to him, so he made a couple of corrections, inputs for the computer. The primary concern was a smooth spot.

Now we were not moving forward, not moving laterally. We were coming down very positively, straight down. I thought we were going to have the smoothest, easiest touchdown that had ever been accomplished on the moon. Our engines began to stir up dust at about 100 feet, which completely obscured the surface. We were landing on instruments. Of course, we had a probe on the landing gear that was searching for the surface. When that probe touched the surface of the moon it would turn on a light in the cabin, telling us to turn off the engine.

BLAST-OFF

The light came on. I called "Contact!" Dave immediately pressed the button to shut the engine—and then we fell. We hit. We hit hard; I said, "BAM!" but it was reported in some of the press accounts as "damn." It was the hardest landing I have ever been in. Then we pitched up and rolled off to the side. It was a tremendous impact with a pitching and rolling motion. Everything rocked around, and I thought all the gear was going to fall off. I was sure something was broken and we might have to go into one of those abort situations. If you pass 45 degrees and are still moving, you have to abort. If the Lunar Module turns over on its side, you can't get back from the moon. . . .

We just froze in position as we waited for the ground to look at all our systems. They had to tell us whether we had a STAY condition.

As soon as we got the STAY, we started powering down. Evidently, we had landed right on the rim of a small crater. Dave and I pounded each other on the shoulder, feeling real relief and gratitude. We had made it.

2

ON THE MOON

"Okay, Houston. The *Falcon* is on the plain at Hadley."

The excitement was overwhelming, but now I could let myself believe it. Dave and I were on the surface of the moon. We looked out across a beautiful little valley with high mountains on three sides of us and the deep gorge of Hadley Rille a mile to the west. The great Apennines were gold and brown in the early morning sunshine. It was like some beautiful little valley in the mountains of Colorado, high above the timberline.

There was the excitement of exploring a place where man had never been before, but the most exciting thing, what really moved me and touched my soul, was that I could feel God's presence there. In the three days of exploration, there were a couple of times when I actually looked up to see the earth—and it was a difficult maneuver in that bulky suit; you had to grab onto something to hold yourself steady and then lean back as far as you could. That beautiful, warm living object looked so fragile, so delicate, that if you touched it with a finger it would crumble and fall apart. Seeing this has to change a man, has to make a man appreciate the creation of God and the love of God.

60

ON THE MOON

There we are in the spacecraft, and even before the dust settles Dave is describing the surface of the moon through the windows of the *Falcon*. "We're sure in a fine place here," Dave tells Houston. "We can see St. George; it looks like it is right over a little rise. I'm sure it's much farther than that," he corrects himself, looking more carefully. "We see Bennett Hill. . . ." But Houston brings us back to immediate business, after encouraging Dave by telling him that the Science Support Team in the back room is doing "slow rolls."

Dave and I have decided that we want to be in the best possible shape while we are working on the surface of the moon, and therefore we are going to sleep before we do any exploring. The doctors don't believe we can relax enough to do it. So, to take the edge off the excitement, we plan to get a good look around before we bed down for the night. We are going to depressurize, open the hatch, and Dave will stand on top of the engine cover. He is scheduled to take pictures and describe the lurain.

Finally, all our homework was done and we were ready to go. We depressurized, and Dave climbed up through the open hatch. As I watched him, I decided he was just a frustrated tank commander. He stood up in that hatch and looked around as if he were the Desert Fox. He described the whole panoramic view; then he began shooting with two Hasselblads, alternately using a 60-mm lens and a 500-mm telephoto lens. He offered me a chance to look out, but my umbilicals weren't long enough and I didn't want to take the time to rearrange them. (The umbilicals lock into your suit at your chest and into the console inside the spacecraft.)

Then we closed the hatch, repressurized the Lunar Module, and began to get ready for our night, though it was still early morning on the moon. We put up the window shades and powered down the *Falcon* as much as we could. The environmental control system and the communications were still operating, but the computers were down, and we turned many of the other instruments off to reduce the electrical load.

Dave was sleeping fore and aft and I was athwart ship, with my hammock slung under his. I noticed that my hammock was bowed out a little bit and my feet were sort of dangling off. It was noisy in the Lunar Module with the pumps and the fans running, something like sleeping in a boiler room. But, man, it was comfortable sleeping! Those hammocks felt like water beds, and we were light as a feather. That first night's sleep was the best I had the three nights we were there.

Houston had us up an hour early the next morning because of a slight oxygen leak. It turned out that we had left a urine dump valve open, and once that was corrected the problem was solved, and we hit the day about twenty-two minutes early.

We got all the helmets and visors aligned and adjusted, and we put each other's boots on and verified that each other's gloves were properly locked on. Dave and I had more conversation while dressing than we'd had altogether in the preceding several days. Then we checked out our backpacks and attached the emergency oxygen supply to the tops of our PLSS's (Portable Life Support System). When I examined mine, I found a large-sized nick in the antenna. We reported it and ended up using that good gray tape to wrap and reinforce it at the weak point so that it wouldn't break off. Finally, when everything was GO, we depressurized the Lunar Module and opened the front hatch.

ON THE MOON

Dave climbed out first. I followed him, but I got hung up in the hatch. Dave talked me out. It's rather a tight squeeze and you are going out backwards, of course, on your belly, and you have to scrunch down to the right elevation and kinda ease the backpack underneath the hatch. As you could see on our lunar television coverage, Dave came down very confidently, but, having shorter legs, I had to reach for each step on the ladder. When my feet came to rest on the footpad of the module, I thought it was the surface of the moon. But as soon as I put all my weight down, the footpad rotated and I had to swing around, hanging onto the ladder, to keep from going on my back.

I could hear Dave's historic landing speech: "As I stand out here in the wonders of the unknown at Hadley, I sort of realize there's a fundamental truth to our nature. Man must explore. And this is exploration at its greatest."

My first thoughts were somewhat less rhetorical. Oh, my golly, I thought, I'm going to fall on my backside in front of all those millions of television viewers.

I just barely caught myself. Recovering my dignity, I heard Dave say, parenthetically, "Well, I see why we are in a tilt. . . . We are on a slope of about ten degrees, and the left rear footpad is probably two feet lower than the right rear footpad." Houston would occasionally refer to our Lunar Module as the Leaning Tower of Pisa—which Dave showed no signs of appreciating.

I looked to the south. I remember saying, "Oh, boy, it's beautiful out here! Reminds me of Sun Valley." The Apennines were very familiar looking. They were rounded and treeless, and there were even mountains that looked like Half Dollar and Dollar Mountains in Sun Valley. They looked like excellent ski slopes.

All around us there was soft material on the surface about three inches deep, just like powder snow. We knew that underneath this dust there might be rocks that we could trip over. If we did, it wouldn't be a serious matter, because we wouldn't fall very hard. I only weighed 26 pounds on the moon, 50 pounds suited up with all my equipment. On the earth, I weighed 160 pounds, the suit weighed 60, and the backpack 80–300 pounds, so it was a real chore to walk. But on the moon, at 1/6 G, it was easy and exhilarating. However, the suit restricted movement. That's why we didn't walk with a natural gait. When you don't have the weight of your legs available to push against the suit, you are constrained as to how far you can move. Consequently, you just use the ball of your foot to push off. That's why we looked like kangaroos when we walked. We flexed the boot and that propelled us forward.

Walking on the moon feels just like walking on a trampoline—the same lightness, the same bouncy feeling. Falling down is the same too. If you fall down on a trampoline, you put your hands out and catch yourself and push yourself back up. You can do the same thing on the moon. The surface, of course, is very soft. The only danger is the possibility of tearing your suit, and that's remote, You'd have to have an extremely sharp object to penetrate the layers of the suit. If that happened, they say your blood would boil. You'd have ten to twenty seconds if the pressure immediately went to zero. If you could get back into the spacecraft and repressurize it before the suit got down from 4 psi to 2.5, you would probably be okay.

Dave and I both fell down twice while we were on the moon. When this happened, we'd help each other up to keep from getting any dustier than we had to; dirt on the suit absorbed heat from the sun and put a strain on our built-in cooling systems.

After a little reconnoitering, we were ready to off-load our lunar dune buggy, the Rover, the first surface transportation ever designed for another planet. We couldn't wait to take a spin.

The Rover was packed in an outside equipment bay in a container that was a little larger than a suitcase. When you look at the Rover it is hard to believe that it will fold up into such a compact package. It is an amazing little vehicle, but then when you pay $8 million for a dune buggy, you expect something pretty good. Even the earth version that we trained in, designed for 1 G, cost $1 million.

With a little help from Houston, we pulled some pins and the Rover sprang out, and then we pulled on lanyards and straps and it assembled itself. The chassis folded into position, the wheels flopped out, and we secured everything with locking pins that slid into catches.

There is no air in the tires; they are made out of woven piano wire stretched to provide a surface like an ordinary tire. The tires are faced with titanium chevrons designed to ride on top of the dust. When you go over a rock, they just bow up and absorb the impact and spring back again. Amazing. We drove over a lot of rocks on the surface of the moon and never broke one of those piano wires.

The Rover is powered by sealed electric motors in its wheel hubs, each one just about the size of the motor in a power drill. It runs on two 36-volt batteries and has a gyroscopic navigational system. It mounts a special communications package that kept us in radio and TV contact with the earth. The whole stripped-down buggy weighs about 455 pounds on earth but only 76 pounds on the moon and is built to carry 2½ times its weight at a maximum speed of about 10 miles per hour. But the Rover felt faster than this. There is no atmosphere on the moon, so there isn't any wind in your face, but there is a great feeling of motion.

Of course, you are constantly dodging rocks and craters. You hit a rock and you are literally airborne. You just bounce into space, float for a while, and then come down. Rover is a flying machine. It is also a great little roadrunner with four-wheel steering—each wheel steers on a different radius. This means that, when you turn right, the right-hand wheel will steer on a smaller radius than the left-hand wheel. The Rover could make a circle within its own ten-foot length. I've never liked safety belts, but we couldn't have done without them on the Rover. It had a definite pitching motion that was a cross between a bucking bronco and an old rowboat on a rough lake—up and down, up and down. You could easily get seasick if you had any problem with motion.

Communications between us and Dr. Joseph Allen, our Capsule Communicator during EVAs on the lunar surface, were so clear that it was hard to believe we were really on the moon. It was as if "Little Joe" were sitting on one of those mountains talking to us. This seemed to bring us closer to home. Actually, the radio signal (S-Band) suffered minimal losses going through space. It was sent out from Honeysuckle Creek, Australia, our prime station. The sending antenna is a 210-foot dish.

Joe gave us minute instructions to cope with each hitch. "Buckle up for safety," he told Dave, for all the world like a parent talking to a child.

Dave drove off while I stayed back at the *Falcon* taking pictures of him with a movie camera, or trying to (the camera jammed). Then Dave's voice to Houston: "I don't have any front steering, Joe."

66

"Press on," Houston answered.

Dave took a little spin in the Rover, and then he picked me up at the front door of the Lunar Module and we took off on Traverse 1, heading directly south. Dave did a remarkable job with rear-wheel steering only, dodging the craters and most of the rocks. We were heading for Hadley Rille and the front of the Apennine Mountains, an area that is most important from a geological point of view. This traverse would give us a sort of security; it would ensure the best possible use of our time in case we had to leave the moon after the first EVA. The traverse was planned to take us directly to the farthest point on the route so that we would have enough supplies in our backpacks and time to walk back to the Lunar Module if the Rover broke down. You always have to have "walk back" capability.

My seat belt turned out to be too short. We didn't realize, when we made the adjustments on earth, that at 1/6 G the suit would balloon more and it would be difficult to compress it enough to fasten the seat belt. So every time we got out of the Rover, Dave had to come around and unfasten my seat belt, and he'd help me buckle up before we could take off. If that wasn't an aggravation. I felt like a baby.

Well, we were bouncing and skimming along, and to our amazement no dust was being thrown up. The big concern had been that we'd be surrounded by a cloud of dust that might keep us from seeing where we were going and prevent us from making any observations. Not so. The fenders worked like a dream in keeping the dust down.

At one point we came over a little rise and there was a crater about twenty feet deep right in front of us. Dave made a quick left turn that threw the vehicle up on the two right wheels. I felt sure we were going to flip. What if the thing did roll over and pin us underneath it? Could we ever release those seat belts so we could get out from under and turn the Rover back over? We never had to find out.

Suddenly Dave shouted, "Hey, you can see the rille!" From the top of this ridge we could look down into and across Hadley Rille.

We were amazed at how huge Hadley Rille was. We could look down about 1,000 feet and across to the far wall at least a mile away. Both Dave and I suspected that Hadley Rille was a fracture in the moon's crust (most scientists had assumed that the sinuous rilles of the moon were in some way volcanic—formed by collapsed lava tubes or the result of flows of volcanic material). Looking to the south along the edge of the rille that faces to the northwest, I could see several large blocks that had rolled downslope three quarters of the way to the bottom. Soon I could see the bottom itself—very smooth, about 200 meters wide, and with two very large boulders right on the surface of the bottom.

"It looks like we could drive down to the bottom over here on this side, doesn't it?" Dave asked hopefully. And he actually wiggled over and found a smooth place that sloped from St. George's Crater into a gully that dropped to the bottom of the rille. "Let's drive down there and sample some rocks."

"Dave, you are free to go ahead. I'll wait right here for you," I told him. I reasoned that we might have made it down and back, but if we had driven to the bottom and something happened to the machine, we'd never have been able to get out. You wouldn't have either the energy or the time to walk up that incline. You'd be completely exhausted, and you'd run out of oxygen. As it happened, the next day, on the second traverse, we learned that the Rover had tremendous hill-climbing capabilities and could climb a 15-degree slope.

ON THE MOON

We drive up on the east rim of Elbow, moving easily up the 10-degree slope, and Houston confirms that we are to stop here and follow the checklist, as planned. This is our first stop. It turns out that we have gained back the twenty minutes we lost. It's a sporty driving course, and Dave has been going like gangbusters. Now we settle down to getting rock samples, with Dave wielding his hammer and me the bag man.

We find and identify basalt, breccia, olivine, and plagioclase. I dig a little trench and get some soil which is quite friable. Dave picks up a rock that looks something like a raindrop, and right behind him I discover one of those fresh craters with a lot of glassy-looking rocks in it. About then we get the signal from Houston to move on up to St. George.

On the way, we produce a constant stream of information on every block, rock, and crater, and we are grading the ejecta from all craters along the way as to how fresh it appears to be. Houston breaks in to ask us if we can see our old tracks as we double back, and we can; the old Hansel and Gretel trick still works.

We took the direct route back because we had already used almost four of our seven hours of the first EVA, and if we didn't scurry back to the Lunar Module we wouldn't have time to deploy the ALSEP, an acronym for our Apollo Lunar Surface Experiments Package. We saw a neat place to go down into the rille on the way back, but Houston broke into our dialogue with "We suggest you don't take that option." We didn't realize how high we had climbed until we looked back. Then we tore down the hill, getting back into the Grand Prix mood again. Just before we hit level, or almost level, ground Dave turned sharply. The front wheels locked and dug in, the rear end broke away, and around we went. We did a 180-degree reverse spin—a perfect christy. Then we crossed our old tracks. It gave me a curious feeling. I knew how Robinson Crusoe felt when he saw the footprints on the beach.

Dave drove over the last knoll and around to our parking position by the equipment bay. Our own little parking lot by our own little campsite—no competition at all. As soon as the cables were straightened out for TV coverage and the gear was unloaded, it was time to put out the ALSEP. We had logged more than six miles in a four-hour Saturday afternoon ride in the Rover.

The ALSEP contained instruments and devices for a number of sophisticated ongoing scientific experiments, including thermal experiments designed to measure the heat flow from inside the moon to its surface and a solar wind experiment that involved the use of an aluminum-foil window shade to catch high-energy particles. I opened the thermal doors of the scientific equipment bay, took out the ALSEP packages, weighing 256 pounds, and attached them to a carrying bar, forming a dumbbell. The carrying bar was flexible. As I bounced along it bounced up and down. It was hard to hold with my hand, so I put the bar in the crook of my arm and jogged out to the selected spot, about 350 feet west of the *Falcon*. We wanted to be far enough away so that when we took off the instruments would not be knocked out by exhaust gases and debris.

When I got out to the site, it was a task taking those packages off. I had to refer to my "cuff checklist," a complete flight plan for surface operation bound into a pad and attached to my cuff. It told us what to do minute by minute. As we finished one page, we turned to the next. The checklist gave me the proper order for unfastening the bolts and precise instructions for removing dust covers, slipping packages out, distances of deployment, and specific orientation.

I had a map that showed where each package went in relation to the Central Station, which is the nerve center that relays data back to earth. The power for this is provided by a radioisotope thermoelectric generator, our little nuclear power plant. It was designed to provide power for years after we left the moon. Erecting the Central Station itself was a key to the whole operation. I recall there were about twenty Boyd bolts around the top platform that you have to release. If they are released properly, the station should spring up when the last bolt comes off. I must have missed one. There were two strings to pull that were attached to cotter pins. Well, I was using a Universal Handling Tool (UHT) to get at one of the strings so I could pull the pins out, and darned if the strings didn't break. So there we were—the Central Station was sitting on the surface of the moon, and those cotter pins were down there and I couldn't get at them.

You can't get down and grovel around on the surface on your hands and knees or you'll get dust on everything. Somehow I had to grasp those pins with my big bulky gloves and try to lift them out. There were several ways of doing it, but time is a real factor on the moon, and we were behind schedule. I didn't have time to experiment.

I prayed to God to show me the right way. The answer was to get down on one knee, holding myself with one hand while I took the pins out with the other hand. This freed the metal tray, which came off and exposed the Boyd bolts. When I released them all, the Central Station erected itself—popped right up, three feet high.

I was still running a little behind schedule, and Dave

was sweating out his heat-flow experiment; he couldn't get the Lunar Drill, a small-scale jackhammer drill, deeper than about three feet. He was supposed to drill ten feet down into the moon to put the thermometer probes in. Two holes were for the heat-flow experiment; he was to dig a third hole to bring out a ten-foot core.

Finally, after a seemingly endless conversation with CapCom about technique, Houston told Dave to take a break and help me. So he came over and deployed the Laser Reflector for me. This involved placing an array of 300 glass reflectors on the surface of the moon so that earthbound astronomers can aim a laser at the moon and fire away at the reflector. When that energy is reflected back it will enable the scientists to measure the distance from the earth to the moon very precisely, within five inches. With a precise reading of that distance they can measure such things as the wobble of the earth's axis. When there is a certain amount of wobble, apparently the forces are redistributed and the surface of the earth moves. Then, the scientists think, we have earthquakes.

After Dave deployed the Laser Reflector, he set up the Solar Wind Experiment. He put up a sail of aluminum foil to catch the electrified particles as they came in; they would actually imbed themselves or they would leave a trace. After three days, we took the foil back with us and sent it to Switzerland. This had been my baby, but Houston told me to knock off and rest, so I talked Dave through the erection and positioning of the screen. Then we got back into the Lunar Module and closed the hatch.

Both of us now realized with some dismay how much our fingers hurt. The pain was excruciating. We took each other's gloves off to see what the problem was. The perspiration poured from our gloves. I looked at my nails but couldn't understand why they were hurting so much; they never hurt on earth. In the six days since launch the fingernails had grown, and they had been immersed in sweat for the last seven hours. The pressure was on the end of each nail. And our gloves fitted tightly against the end of the fingers so we could have some feel through the heavy material.

I took my scissors and cut my nails back just as far as I could. From then on, on the next EVA, there was no problem at all. I said, "Dave, cut your fingernails—that will solve the problem." But Dave would not cut them. He just put up with the pain through the next two EVAs.

I was really dragging. Tired, exhausted, hungry, thirsty. I would never have gotten through that day without a little nourishment. We had food sticks that were about twelve inches long secured by Velcro at the neck ring and extending up into our helmets. The stick was positioned so that we could reach down, grab it with our mouths, and take a bite out of it. I had gobbled mine down after the traverse, and it had given me energy to put up the ALSEP.

I didn't get any water during the EVA. Each of us had a plastic bag full of water attached inside our suits. There was a nozzle that you'd bend down to open a valve so you could suck the water out and drink it within the protection of the space suit. But I never could get my water bag to work—and I never got a single drink of water during the whole time I was out on the surface of the moon. This plus heavy perspiration probably contributed to my losing all the potassium in my system. This was to cause me trouble later.

From the moment we got the Lunar Module repressurized that night and got out of our helmets, we were overwhelmed by a strong acrid smell like gunpowder. This was the lunar material that covered us, a fine black dust, being exposed to oxygen for the first time. As it oxidized it gave off that strong odor.

Then, as we were struggling to get the dirt under control, Houston reported encouraging news—that Al Worden's scientific package was transmitting an abundance of data.

Evidently, one of us had bumped into the water gun when we got back into the module that afternoon. Later we found tell-tale water on the floor; the water gun had been broken off. Upon closer examination, it turned out to be the bacteria filter that was broken. By taking the filter off, we could replace the gun and stop the leak.

We really guzzled down the water that night. I had been seven hours without a drink and sweating like a jogger.

You can imagine how dirty the Lunar Module was. We took all the rocks in there, and we had this graphitelike dust all over the floor. We took off our suits, trying to make sure we didn't get dirt inside them. We stowed them in the back end of the Lunar Module, where they would be out of the way. We put the gloves on the suits to protect those rings, so we would have the proper seal the next time.

We debriefed, we planned the next day's EVA, and finally we bedded down for our second night on the moon. I didn't sleep nearly as well as the night before. It could have been the pain in my fingers. Dave and I talked briefly about the day, and he seemed to think it was a successful day except for the Lunar Drill.

The temperature at 1/6 G was comfortable but we couldn't block the light out completely. I had a horrible headache, probably because I hadn't pulled the sun visor down on my helmet. So I took two aspirin—the first and only medication I took on the entire flight.

Houston woke us up the next morning at 2:35 A.M.

Houston time. We started with a problem. According to the data Houston had received during the night, the Lunar Module had lost 25 pounds of water since the EVA the day before. They suggested we look around and see if we could find any signs of this substantial load of water. Gordon Fullerton, spacecraft communicator, suggested that it might be behind the ascent engine cover because of the angle of the spacecraft. Gordon was right on the marker. The water was there, and we scooped it up. It was a shock to see all that water back there—near some electrical connections.

After we had scooped the water into lithium hydroxide canisters, we suited up, depressurized the vehicle, opened the hatch, and threw the containers out on the surface of the moon. We had been eating breakfast at the same time we were cleaning up the water, and now it was time to go on EVA 2. We were eager to get out. The painful physical work with our hands was behind us, and today we were to look for rocks.

The mountains of the moon were our primary objective. We hoped to find some light-colored material that the scientists had told us existed up there. This is the day we had a cross-checklist on each wrist—we were really programmed to the hilt.

After some simple instructions from Houston, the Rover's front-wheel steering began to work like magic. It was a scenic trip past a number of major landmarks like Index, Salyut, Earthlight, Domingo, and Dune craters. The going was rough, but it suddenly smoothed out as we drove onto high ground near the Apennine Front. We could look ahead and see craters splattered right up the slope of Mount Hadley Delta. The driving improved as we moved along the front. "Boy, that's a big mountain!" Dave exclaimed, when we got Hadley Delta framed squarely ahead of us. The soaring height of the mountains of the moon, from the base up, seemed to us to compare with Everest and other great mountains of the earth.

Making the most of our powder base we did a couple of christies and headed right for Spur Crater, where we parked Rover.

Spur Crater was a real gold mine. As soon as we got there, we could look over and see some of this white rock. Immediately, I saw white, I saw light green, and I saw brown. But there was one piece of white rock that looked different from any of the others. We didn't rush over to it; we went about our job the usual way. First I took down-sun shots and a locator shot about 45 degrees from the sun line, and Dave took a couple of cross-sun shots.

Our sampling routine was automatic. I'd let Dave make the decision as to which rock we were going to sample. Then I'd place the gnomon on it—this is a device that tells us what the local vertical is and also has a color chart for color references.

Finally, we worked our way over to the white rock that had riveted our attention earlier. It was lifted up on a pedestal. The base was a dirty old rock covered with lots of dust that sat there by itself, almost like an outstretched hand. Sitting on top of it was a white rock almost free of dust. From four feet away I could see unique long crystals with parallel lines, forming striations. This was exactly the kind of rock we were looking for; it confirmed the suspicion that the mountains of the moon were made from rock, lighter in color and lighter in density.

If the moon were composed entirely of the sort of dense rock that had been sampled on previous missions, the moon mass would be much greater than the scientists had calculated. They knew there had to be a lighter material on the moon, and they speculated that this would be anorthosite or plagioclase.

I think it was providential that this particular rock was lifted up and displayed to us. We brought it back. Scientists will be analyzing it for years to come, but the University of New York at Stony Brook has age-dated it at 4.15 billion years, plus or minus .25 billion years. This confirms the fact that the earth and the moon were created about the same time. The oldest thing on earth discovered thus far is about 3.3 billion years old, so this little bit of ancient history is another clue to the creation of our universe. After we returned, we found that our treasure had been named the Genesis Rock. I think this discovery alone would have made our mission a success.

After we bagged the Genesis Rock and collected a good number of fragments, from football size down to walnut size, I got a bag of soil from the immediate vicinity and we loaded up. Houston was on our back to get rolling.

When we got back to the *Falcon*, I got a bunch of photographic assignments from Houston that included panning the Lunar Module site and photographing the descent engine and the Solar Wind Experiment. Dave was asked to try the Lunar Drill again. He had already damaged his fingernails to the point where several of them would turn black by the time we got back to earth, and he had damaged his temper considerably too. He told Houston that he hoped fooling around with that drill was more important than studying the geology of the area, but then he decided to give it the old college try. Subsequently he completed the heat-flow experiment and finally reached the required depth. But the trouble was—he couldn't get it out. I was asked to start on my Station 8 ditching experiment.

If you think digging a ditch is dog's work on the earth,

try digging a ditch on the moon. The big limitation is the suit and the fact that you are clumsy at 1/6 G. I had practiced on earth and come up with the technique that most dogs use. You spread your legs and push the dirt between them. I solved a dog's job with a dog's technique. This method worked perfectly on the moon. I had a scoop that was about three inches wide and four inches long, not a very satisfactory digging tool.

First I went down through a layer of fine dark-gray material, like talc or graphite. Then I started to encounter coarser soil that was darker in color. At 12 inches I encountered a very resistant layer of soil like hardpan; it almost looked moist. I finally managed to scrape that layer clean and get the required dimensions. Then we did the penetrometer test, to measure the cohesiveness of the soil, and took pictures.

We had found that the depth of the dust on the surface of the moon varies from a few inches to six to eight inches. The pulverization of the soil is probably caused by the impact of meteorites striking the surface of the moon over billions of years. These meteorites vary in size from micrometeorites the size of a pinhead to projectiles fifteen miles in diameter.

Just after Apollo 16 left the surface of the moon, the seismic device detected the impact of an object estimated to be ten feet in diameter. It hit very close to the scientific base and created a crater about the size of a football field. So large objects are still striking the moon. The earth is protected from this by our layer of atmosphere. The shooting stars that you see at night are incoming freight that would strike the earth if they didn't burn up when they hit the protective belt of our atmosphere.

We are breaking out the TV so Houston can catch the flag-raising scene in which Dave and I will star in a few minutes. I have the flag stashed nearby, waiting for Houston's signal.

"Dave, we've got a lot of time; we're going to deploy the flag now and we need the TV, please."

We don't want to stop work, we just want to stop hearing from Houston for a while. So we go about our busy work talking to each other, going nice and slow and easy. We've even got the TV equipment all dusted and the camera set up. And then CapCom says, "Jim, at your leisure we'd like for

to deploy the American flag, please." We pick the spot, with Mount Hadley well placed in the background. And I push the staff in and hit it a couple of times so it'll stay up for a few million years. Dave and I salute the flag. It's beautiful.

We finished our yard work with time to spare. Then we got in the Lunar Module and closed the front hatch behind us. We pressurized the vehicle and gratefully got out of those suits—what a relief! And with a little help from Houston we recharged our PLSS's. Then, after a while, good old Alfredo passed over in the Command Module, and we really had a chat.

"We got a load of about a hundred pounds [of rocks] today. Got up on the side of the mountain and got a good look around," we told Al. "Things are going real well. Hope you can see those Rover tracks, because outside the LM here it looks like a freeway."

Al promised to make a nice place for our rocks. Then I asked him to throw my soap down. We sounded like a bunch of campers yelling at each other across a canyon.

Then we said good night to Houston, and they faithfully promised to wake us up in seven hours. Yet we would not have even seven hours with a call set for 2:00 A.M. Houston time. This aggravated me, because I could see I wasn't going to get a good night's sleep. As we lay there, we were thinking about the fact that our EVA the next day was going to be a little short. For us, it really represented our most important EVA because we were going to visit the canyon. But we were also going to the Northern Complex, where we thought there would be clear evidence of volcanism.

I was happy about what we had accomplished that day. I was very grateful when I thought about this providential care over us. The Rover's steering was working fine, and we had found the Genesis Rock—certainly God had been watching over us that day.

I prayed on the moon at the same time I pray on earth— night and morning. I remember thanking God that night for the guidance He had provided, and as I talked to Him there was a sense of His nearness.

Dave and I never talked about this aspect. In fact, Dave and I didn't have any conversation at night at all. We realized it was important that both of us go to sleep. And frankly, we both snore, and I guess I snore louder than Dave. So there's a contest—each of us wants to drop off before the other starts snoring. That night, just before I went to sleep, I asked the Lord to give me the right time to quote a verse of scripture that had been significant to me throughout my life.

After I prayed, I thought back over the day. We had been driving back to the *Falcon* that afternoon, really exhilarated by the beautiful mountains on three sides of us. It was the inspiration of the mountains that made me think of trying to talk Dave into having a brief religious service. With the light hitting Mount Hadley we could see the parallel lines in sharp relief. This pattern of layering was puzzling to the scientific community because it raised geological questions that were difficult to answer, but the beauty of this mountain had moved me in a different way. I wanted to hold a religious service on the moon, but Dave wasn't too enthusiastic. It would mean getting authorization from NASA headquarters.

It turned out that we did not have time. If we had done it anyway, we would have been under the gun, so to speak; we would have been criticized. Well, I prayed for the right opportunity, and then I went to sleep.

We were awakened at 12:37 A.M. Houston time and got on with discussing the traverse plan for that day. "Basically," Houston said, "the EVA is going to last somewhere between four and five hours, so it will be a short EVA. . . . From here on out, it's gravy all the way and we're just going to play it cool, take it easy, and see some interesting geology. It should be a most enjoyable day."

But we know the heat is on. The time is short for us, the pressure will build to make the lift-off deadline, and we suspect that we will be asked to fiddle around with the Lunar Drill again. We get the word from Joe. "We're going to ask you to stop first at the ALSEP site and spend a few minutes recovering the successfully drilled core tube." Of course, we are eager to get up to the Northern Complex. We don't want to lose traverse time.

"We're going to ask you, Jim, to help Dave with the core removal," Joe added.

"Good," Dave says. "That shouldn't take him more than a half hour. I might just let him do the whole thing. He needs some experience."

When we got to the site, Joe said quietly that Houston was standing by for the TV remote. The last thing we were interested in was live TV coverage of us struggling with that drill.

It was rough using the drill and trying to listen to Joe's suggestions from Houston at the same time; we had to stop for a second and then start up again. "If we get hung up, we'll let you know," Dave told Houston, "and we'll give you progress [reports]."

As we got down to business, my relationship to that drill changed abruptly.

"I don't think it's worth doing, Jim," Dave said. "We're not going to get it out."

"Dave," I said, "we're going to do this. We're going to get this drill out."

Dave and I had our shoulders under the drill, and I counted: *"Get ready, get set, and heave."* Well, Dave was out of sync. I was heaving and then he was heaving.

Finally, I gave up and said, "Dave, you call the numbers. I'll heave when you tell me to." So we did it that way. Both of us strained with all our might, trying to lift that thing up, and finally it started breaking. Man, oh, man, did we get it up! We almost flew off the surface when it came out.

I knew we would get that thing out. I was going to stay there and get it up no matter what it cost, once I got involved in it.

It was essentially a ten-foot section of pipe with a core inside. The core itself looked like soil with some coarse black and brown fragments in it. We had to try to unscrew the hollow sections of the drill so we could break the core into manageable lengths and store them in the Lunar Module. Frankly, we were unable to do it except at one point; we were able to divide the core into only two parts, each about five feet in length. But Houston was satisfied, and so were we.

Soon we were driving west in our trusty Rover, at last

on the long-delayed traverse. After crossing lurain that has the aspect of rolling sand dunes, we came upon a striking view of the far side of the rille. We parked near the rim of the canyon. We both marveled at the layering in the far wall.

These patterns may represent successive lava flows that formed the surface of the basin of the lunar "sea." There were shadings of gray into brown, different textures and patterns. Remember, this canyon is 1,000 feet deep and about 4,000 feet from rim to rim. CapCom nervously asked us how close we were to the edge of the rille. Actually, we were about 50 meters away.

Dave and I agreed in our theories of how Hadley Rille had been formed; it probably was a fracture (possibly a moon quake) that had caused the moon's surface to spread like a cooling joint. But since the scientists have studied the pictures, the most popular theory is that Hadley Rille was probably a lava tube that collapsed.

After we started rolling again both of us began to feel the heat; the sun was getting higher. At the end of our third earth day, the moon day was warming up. We were barreling along and I told Dave to take it easy; "We want to get a good trade-in on Rover." We stopped and examined one angular fragment of rock that amazed Houston. It was about four by five feet with very large vesicles of about two to three inches in diameter facing the southwest. I think CapCom was afraid we were going to try to get it on the Rover. We were being hustled along now, and as we drove back I got the impact of the sun again; it was really fierce.

Now comes the chance I have been waiting for. Dave has a lyrical moment. "Oh, look at the mountains . . . when they are all sunlit. Isn't that beautiful?"

"Dave, that reminds me of a favorite biblical passage in the Psalms: 'I will lift up mine eyes unto the hills, from whence cometh my help.' But of course we get quite a bit from Houston, too." I just tuck it in swiftly, but I feel better after that.

We came upon our tracks and saw the Lunar Module. We didn't get a chance to go up to the Northern Complex to see Pluton, Eagle Crest, or all the other good places. Instead we drove right back to the ALSEP, and both of us went to work cleaning things up. CapCom was telling us to break down the core stem first. This time, I despaired of ever being able to break it down further. "The crew will break down; the stem never will," I said. But Dave made up his mind that he would break it, and he took that drill apart into three sections. The heat was really getting to us by now, and we both felt we could drink about three gallons of Gatorade.

We stripped the Rover of everything that was valuable —cameras, gear, rocks, and tools. It was Dave's job to drive the Rover about 300 feet east and position it so that the TV camera could be focused on the Lunar Module to record the lift-off. The TV camera would continue to run on the battery power of the Rover. The ground could power it off, bring it back on, do whatever they wanted with it—that is, until it got caught in one of those stops. We had just barely broken in the Rover, and the batteries had a lot of hours left on them. (They are there now, Rover and the TV, in that little used-car lot up there, awaiting the next visitors to Hadley Base.)

The TV camera was directed right at Dave, and he brought up a subject that would later have fateful consequences for all of us. "Okay," Dave said, "to show that our good postal service delivers any place in the universe, I have the pleasant task of canceling, here on the moon, the first stamp of a new issue dedicated to commemorate United States achievements in space. I'm sure a lot of people have seen pictures of the stamp. The first one here on an envelope. At the bottom it says 'United States in Space, a decade of achievement,' and I'm very proud to have the opportunity here to play postman. Pull out a cancellation device. Cancel this stamp. It is August the second, 1971, first day of issue. What can be a better place to cancel a stamp than right here at Hadley Rille!"

Then, while we had the attention of our audience, Dave and I did what we called "The Galileo Experiment." It involved Dave dropping a falcon's feather from one hand and his geology hammer from the other. Sure enough, hammer and feather floated down side by side and hit the ground at the same time, about 1.3 seconds. Actually Dave had two feathers that he had taken from the falcon mascot at the Air Force Academy. I played a role in this experiment, too. I accidentally stepped on the feather.

"Where's my feather?" Dave asked. He apparently wanted to take that feather back, and he was perturbed with me. When we got back to the module, he said, "Jim, why did you step on my feather?" All I could say was, "I'm sorry, Dave, I didn't know you wanted to bring that feather back." We could only find my two big footprints. I'm wondering if hundreds of years from now somebody will find a falcon's feather under a layer of dust on the surface of the moon and speculate on what strange creature blew in there.

Dave had already put out the plaque commemorating the astronauts and cosmonauts who had died in space exploration. I had previously been responsible for getting the names on the plaque in alphabetical order, and seeing that the

Russian names were spelled right. Al had provided the statue. While Dave was involved with winding up his chores, I found that unaccountably there were no more activities planned for me. With fifteen or twenty minutes with nothing to do, I actually had a small vacation on the moon.

I started running around the Lunar Module in circles, and I did some broad jumping. Just having a ball—you know, like a little kid. Even in the space suit, I could broad jump about ten feet, about three or four feet in the air. No telling how far I could have jumped if I hadn't had that suit on. It was the most relaxed time I had on the surface.

Houston was urging us to get into the Lunar Module. It was time to come home. "Move the baggage into the *Falcon* and climb in." CapCom got a bit lyrical. "As the space poet Rhysling would say, we're ready for you to come again to the homes of men on the cool green hills of Earth."

Then CapCom told us the mission had been accomplished: "Dave and Jim, I have noticed a very slight smile on the face of the professor. I think you very well may have passed your final exam."

As we began to get squared away, Joe Allen came on and told us that he had enjoyed it. Once he was off our backs and patting us on our shoulders, we began to feel slightly euphoric.

We had plenty of housework to do before lift-off. We had to stow the rocks and soil in bags and other containers. The cameras and scientific instruments had to be accessible according to how likely they were to be used on the flight home. Since we were taking 250 pounds of moon rocks and soil back with us, we had to dump everything that we wouldn't need. We carried a kind of fish scale that we used to weigh everything we brought into the spacecraft. We left the steps behind because they were part of the descent stage, and we pushed out the backpacks. We kept the emergency oxygen system in case we had to use it to do a vacuum transfer from the Lunar Module to the Command Module. It was good to get rid of our big trash bag—the remnants of all our meals on the flight. I inadvertently left a few used defecation bags in that trash bag too, so they are lying on the surface at Hadley Base. Our dirty campsite will be there for a million years.

We taped caps over the ends of the core tubes, taped them together, and then secured them to the aft portion of the floor of the Lunar Module. By now, this core had for us the highest priority of any item that we were bringing back. It was clearly our core. About this time the water quantity light came on, indicating that we had exhausted all the descent water on board. We had hit it right on the money. It was time to power the Lunar Module up again.

In simulations we would always have an abundance of time so that we could carefully recheck and talk things over with each other, but here in the real world on the real moon we were faced with the fact that we were behind time. Of course when we trained on earth, we had our space suits on and we worked in the same environment, but at 1 G as opposed to 1/6 G. One sixth makes it much easier, but we had not anticipated how much time it would take to bring the samples off the surface of the moon. For example, it took 400 percent longer than we thought it would to get that core. Yet we were thinking clearly and working together as a team very efficiently, and there was this feeling of mental power that I mentioned earlier.

At this point, we were not tired, since we had only been out on the surface for about five hours on our last EVA. I wish they had delayed lift-off another revolution. It would have given us the two hours we needed to explore the Northern Complex. I think, looking back now, that it would

87

have been wise to stay another two hours. But at about 12:07 we got the GO for lift-off, and the ascent stage engine fired up automatically and we were up. It almost sounded like the wind whistling past us. What a view of the rille; the Rover tracks. . . . Hey, it's a good smooth ride!

3

FLYING HOME

CapCom: *"Falcon,* Houston, you're looking good at three minutes."

One moment you are on the moon, looking out over the site. The next instant you are fifty to a hundred feet in space, looking almost straight down at the place where you camped for three days. When we pitched over we could see the descent stage, the tracks we made, and the scientific base.

About ten seconds after lift-off we were looking down when suddenly we heard the Air Force song: "Off we go, into the wild blue yonder. . . ." This was kind of surprising, because Dave had briefed Al to turn on that music at one minute *after* lift-off (that first minute is rather critical) but here it came in at ten seconds. It really caused some consternation in Houston. First, they thought somebody was playing a trick in Mission Control, so they conducted a big search. They asked for radio silence—it was a tense situation. Finally, they realized that probably we had turned it on.

Shortly after lift-off we did a yaw maneuver to ensure good communications, and this placed the rille right in the middle of the window so we could get some more good pictures. We continued to accelerate until we reached 5,000 feet per second, at which point the engine shut down. We were in orbit around the moon in a 9-by-45 orbit.

TO RULE THE NIGHT

Looking back, I had expected to have a little concern about lift-off. If that one ascent engine does not light for some reason or other, you are there forever. There is no chance of the spaceship picking you up. The Command Module is not designed to land; it is designed to reenter the atmosphere and splash down in the ocean. If it were designed to land on the moon, it would be a completely different vehicle, like another Lunar Module. However, there are many ways of ensuring the ignition of the one ascent engine, and the engine itself is well protected.

As it happened, I didn't have time to be concerned about lift-off. We came into it so fast that it happened before I could think about it.

Shortly after insertion in lunar orbit, we powered up the rendezvous radar and put it in a search mode, and then our radar beam locked onto the Command Module. We knew we'd got lock-on because the range meter started moving; it tells us how far we are from the Command Module. After a little while we could actually see the tracking light on the *Endeavour*. Al was pointing at us, and the transponder in his spacecraft responded to our radar pulse and sent a message back. So in addition to having information from the radar, we also had information from VHF, a radio device, as a backup.

Our engine had shut down, and we were weightless again. As soon as we achieved weightlessness, all the lunar dust started to float up. I was glad we had our helmets on to protect our eyes, because I saw all our old adversaries: the particles of glass from the broken tape meter were coming out of the woodwork, lunar dust was everywhere, and all sorts of crud were circling us.

Al must have been forty to fifty miles away when we first saw his light, and then we crossed the terminator into bright light and could see him getting closer and closer. We were coming in lower, so we were gaining on him. When we got to about a hundred feet, we slowed down our rate of closure and maintained a hundred-foot separation. This was a great opportunity for us to take pictures of the Command and Service Module. With a little maneuvering we got into position to take pictures of the Scientific Instrument Module (SIM) Bay on the *Endeavour*, which was the first spacecraft so equipped. Then we went into position right in front of Al and held it.

Then we engaged in high style—a good hard docking. There was a kind of thud as the two modules came together. I could hear the capsule latches grab, so I knew we were mated up. Al opened his hatch; we opened ours. "Welcome home," Al said.

A storm of debris moved toward him. Naturally, he objected. Here he has a beautiful clean spacecraft, and our pollution drifts into it as we open up. We tried to keep our hatch closed unless we were physically transferring something, but we had to move all those boxes and bags of rocks into the *Endeavour*.

For the first time in the flight I felt comfortable in a weightless condition. Going from 1/6 G to zero was an excellent transition. I was prepared. I could tumble; I could move my head and my body rapidly; I was completely at home.

Dave and I wanted a few personal souvenirs, so we took the utility lights and other loose parts off the Lunar Module. We were behind schedule, so we were hurrying. We knew we had to jettison the Lunar Module at a certain time so that Houston could fire its engine by remote control and crash our faithful LM into the surface of the moon.

We were slightly rattled by the time squeeze, but we

thought we had everything that was to be taken from the Lunar Module. I was the last one to leave. I worked my way down the checklist of switches that had to be set in the proper positions for the last burn. By this time we had powered the *Falcon* down and were getting our oxygen from the *Endeavour*.

We went through our undocking procedure after the hatches were closed and were assured we had a good seal on the hatch. Then we heard, "Fifteen seconds to LM jettison. Hope you let her go gently; she was a nice one." We really felt this separation; she was the best. Our LM just sat there in space as we backed away. Then the ground sent a signal commanding the engine on, which slowed the *Falcon* down so much it left orbit and went into a carefully calculated trajectory. We were hoping that we might be able to see it hit the surface, but we couldn't. We were in darkness again, and we had our suits on, but our helmets and gloves were off. All of a sudden, for the first time in the flight, I was really tired.

I told Al and Dave, "Just let me lie here for a little bit. I am physically exhausted." I lay down for about five or ten minutes as we were getting ready to take off our suits and prepare for bed. About that time, we got an unexpected call from Deke Slayton.

"You guys have had a hard day," Deke said. "I suggest that all of you take a sleeping pill this evening."

"We hear you, Deke," I said, laughing. "We have had a busy day." I thought he was joking too.

Well, it turns out that they were looking at heart rates, and this is when they detected that I had a bigeminy rhythm. With this condition, the impulse for the heart to contract comes from two sides of the organ at one time, and the heart gets confused. It often is caused by severe fatigue.

FLYING HOME

Our conditions were evidently aggravated by a drastic loss of potassium. But we didn't take any medication.

After nine hours of fairly good sleep, we woke up at 8:15. I felt normal despite my heart problem. We settled into lunar orbit, using the scientific equipment and making observations of the surface of the moon. We were all working with the SIM Bay now. Up to this time, Al had been operating all this equipment on his own, and he did a great job. Now, with all three of us guys there, Houston decided to change the flight plan. Each morning, when we awoke, there would be a whole new page of instructions waiting for us. It became very confusing. Dave and I didn't understand this complicated equipment nearly as well as Al did, and I think we achieved less than perfection.

The first day in orbit, which was August 3, we had one more query about the core stems that had preoccupied Houston. CapCom asked, "Is there a three-unit segment of those core stems some place in that Command Module?" "Joe, we wouldn't lose sight of those cores for all of the tea in China—that's number-one priority."

Houston read me a telegram that had come in from my mother, dad, and brother. They reminisced about our trip together to the top of Mount Whitney and said that they were proud of me and were with me in spirit on the moon. Getting a telegram from the earth was a great morale booster.

With the high-resolution cameras in the scientific bay, we took pictures of the moon that turned out to be extraordinary. From sixty miles up, we got photographs of the surface that revealed the Rover sitting next to the descent stage—this means we were resolving a six-foot object sixty miles away. (This sort of high-resolution photography was used to select the Apollo 17 landing site. Al's observation of the Littrow area revealed what he thought were cinder cones,

93

and because of the intriguing details of these formations, the Apollo 17 crew landed in this area and explored it.)

Evidently Dave had pulled a muscle in his back when he was struggling to get the core tube out. He must have been in intense pain, but he never mentioned it. He didn't want anything to mar the flight, and resented any suggestion that any of us were not in top physical condition at all times.

That night before we went to bed, CapCom asked us whether we had actually taken the Seconal the night before. We admitted we hadn't. CapCom urged all three of us to take a pill that night. Although we said that we thought it was unnecessary, they pressed their recommendation.

At some point when we were sleeping, Houston noticed some irregularities in Dave's heartbeat. If I had been the only one to exhibit these effects they would have thought it was an individual reaction, but when Dave began showing the same signs they really suspected that something had happened to all of us. Looking back, I think we both strained too hard on the drill.

They woke us up about 4 A.M. with what Houston called "a message from Richard Strauss, Arthur C. Clarke, and Stanley Kubrick." As usual Karl Henize was full of surprises; he had piped us a brief passage from the science-fiction movie 2001: Kubrick's movie, Clarke's book, and Strauss' tone poem. At the end of our newscast that morning, the boys in Houston struck a homey note with a report of continuing rain and the comment, "Going to be a lot of grass-cutting to do when you get back down here, guys."

The temperature in the Command Module was so high when we were on the sunlit side of the moon that we were uncomfortable even in our underwear. Outside it would be about 250 degrees F. Then, when we went on the dark side of the moon, it would cool off rapidly until we were quite comfortable. On the bright side you could put your hand against the window glass and feel heat that could cook an egg. Many times I just covered the window, even though it was beautiful to look out. I didn't want the heat coming into our spacecraft. We stayed in our long drawers all the way home.

At some point after we were in lunar orbit, Dave and I were going over the list of things that we should have taken from the *Falcon*. "Jim," he said, "did you bring over the PPKs [Pilot's Preference Kits]?" I said no, I thought he had; Dave had been cleaning out that part of the LM. Well, Dave had not brought them over, and it became apparent to both of us that we had left many valuable things in the Lunar Module.

We had both brought along many items for ourselves and for friends that would have been priceless souvenirs from the moon. There were envelopes, medallions, stamps, medals, flags, shamrocks, and coins. Dave and I had at least a hundred two-dollar bills that we were going to split after the trip. Since I am Irish and was born on St. Patrick's Day, I had planned from the time I was first selected for the program in 1966 to take shamrocks to the moon.

We had had some of Al's things, too, and I had carried dozens of things for associates back at the Manned Spacecraft Center in Houston. One friend had given me his wedding ring. I had a hard time telling him that it was gone, that his marriage would be perpetually celebrated below the lunar surface, deeper than Dave's drill hole. Astronaut Walt Cunningham, who had been on Apollo 7, told me he had always meant to carry a Marine flag on his flight and had overlooked it, so he asked if I would carry one on our flight and bring it back to him. The flag was also in that PPK. When we got back I had to tell Walt what had happened.

"Walt," I said, "don't feel too bad. You probably own

95

the first Marine Corps flag to be planted on the moon." (Talk about "planted," it was really deep, the way that Lunar Module dug into the surface.) I hope that somebody will visit that crater some day and bring back our PPKs.

There were a number of things we left on the moon purposely. I left some medallions, flat pieces of silver with the fingerprints of Mary and our children. And as a result of a letter that I got two months before launch, I also left a small portrait of J. B. Irwin. A young lady sent me a picture of her father, J. B. Irwin, saying that he had talked about his desire to go to the moon all his life. He died at seventy-five, before the first manned landing. I thought it would be a gracious gesture to take J. B.'s picture and leave it on the moon.

It's up there. But unfortunately I threw away the envelope with the address, so I have never been able to tell the lady that her father is sitting on the moon where he always wanted to be.

The second day in orbit, at about 2 P.M., we were scheduled to have a "shaping burn"; it would last 3 seconds and make it possible for us to put out a subsatellite with a year's lifetime.

We spun the subsatellite out just as our spacecraft was crossing the lunar equator on its way to the apolune—it was jettisoned to the north to keep the spin axis perpendicular to the sun's line of sight. This put maximum sunlight on the solar cell to keep our baby working. We watched the characteristics of the spin and noticed an oscillation of about 10 degrees. Everybody was happy—we were leaving a little friend behind, and we were working toward the Trans-Earth Injection (TEI) Burn behind the moon that would head us back to the earth.

At about 3:54 P.M., Houston was on the horn saying that we were GO for trans-earth injection. "Set your sails for home. We're predicting good weather and a strong tail wind, and we'll be waiting on the docks."

Back in Houston at the Manned Spacecraft Center the gang was all together. Our *Endeavour* was eighteen minutes away from trans-earth injection, and the entire backup crew was at the CapCom console in the Mission Control Center. Deke Slayton, our boss and the Director of Flight Crew Operations, was there, along with his deputy, Col. Tom Stafford. Everybody was waiting for contact.

As we rounded the corner of the moon, Dave radioed: "Hello, Houston, *Endeavour* is on the way home." And so we were; we had an almost perfect burn. We were targeted for a splashdown location 285 miles north of Hawaii; the time would be approximately 295 hours after blast-off. The only problem should be "corridor control"; that is, putting the spacecraft in the proper entry corridor to get the exact angle of entry so that we wouldn't either skip right off the atmosphere and back out into space, or burn up like a shooting star. We were aiming for a very small cone, plus or minus half a degree.

The burn is so long you can't help but wonder if there is going to be enough fuel on board to complete it. As it ended, I think we had 5 percent fuel left. That burn is usually so precise you could go all the way back to the earth without firing again. It doesn't jolt you or slam you back into your couch—it's smooth, but it is very positive acceleration, much greater than coming into lunar orbit. We are stripped down—the weight of the Lunar Module is gone, and that means quite a bit.

At one point, Dick Gordon, who had been Command Module Pilot on Apollo 12, said, "And could we pry out of you guys any comments on the moon as you leave?"

We replied: "We are right on the terminator. It shows

very distinctly all the topographic highs and lows. And we can see some major rilles and circumferential rilles extending from the central peaks."

Gordon, backup commander for our flight, was obviously getting a little nostalgic with this description. Once he triggered us we kept firing. We told him we were overcome by the spectacle of the brilliantly lighted moon from space. Remember, when we had cruised in to the moon, it was a dark mass that lighted blindingly only when we got down over the surface. Now we had the same breathtaking views of the moon going away that we had had of the earth when we were lunar-bound.

We had a certain amount of disorientation, too. South was up, and we were looking right up and down the terminator, so in effect we were upside down looking at new territory that we hadn't seen during our orbits. When I looked out of window 5, my window, I was looking north, and I got a halfmoon view. You could see the moon all in one gulp! At the terminator it was this striking gunmetal gray, shading out almost to white at the subsolar points. The whole mass had this dramatic, unreal look.

That evening, Houston took the night off because of a weak signal on the land line to Madrid. Once air-to-ground communications were shut down, they would correct the situation.

"Houston is out for the night," CapCom said.

"Don't go too far out," we said, almost wistfully. When you are out there in space, Houston is the lifeline.

We went to bed that night feeling like a bunch of kids whose parents had gone out for the evening. But Houston was right there, bright and early the next morning.

A little after 6:30 A.M., CapCom reported that we were

leaving the lunar sphere of influence. "It's all downhill from here on in." The displays in Mission Control having to do with velocity and distance switched over at that moment from lunar to earth reference. They reckoned that we were 177,225 nautical miles out from the earth and approaching at a velocity of 2,855 feet a second.

This was Al's big moment. He was scheduled to star in today's EVA, a space walk. He would go out the open hatch and along the outside of the spacecraft back to the service module, hand over hand, to retrieve film cassettes from the mapping and panorama cameras. It was necessary to get the film from the SIM Bay before we encountered the terrific heat—5,000 degrees F.—of reentry. The plan was to depressurize the spacecraft at about 10:24 A.M. and get Al out into space twenty or thirty minutes later.

We were weightless, and when we opened the hatch it was just like a vacuum cleaner pulling all the loose stuff from the inside out into space. Everything started floating out. My toothbrush floated by; it had been in hiding. A camera came by; one of us grabbed it. We were all leaping around, trying to catch the important stuff. Then we got the hatch open wide enough for Al to move outside. Meanwhile we were getting the TV camera ready to record the EVA. Finally we got it aimed right, and Houston got the picture of the man in space.

Actually, I goofed up. When I hooked my suit umbilicals, I wrapped them the wrong way around the strut. I realized this late in the game, but I didn't want to hold up the operation, so I didn't mention it. This gave me less freedom of movement—I couldn't get out of the hatch quite as far as I should have been able to. I had to force my hand out to reach the movie camera (attached to a boom) to turn it on. I just struggled against the button. I saw the green light and thought the camera was on, but it wasn't working.

There I was. I didn't have enough length to take good pictures of Al. I also had a safety tether on that was attached to the Command Module to take the strain off of the umbilicals. Of course Al had umbilicals and a tether. He also had an emergency backpack so that if the umbilicals failed he could actually get emergency oxygen for about thirty minutes.

Al was skimming along through space at the same velocity as the Command Module—when he pushed off from it, he increased his velocity imperceptibly. If he became separated from his umbilicals and from his tether and was free-floating, theoretically we would have to maneuver the spacecraft to reunite with him.

Al made one quick trip and got the pan camera safely inside. Next time he brought in the mapping camera cassette.

It was a beautiful sight as I looked out there: absolutely black, but there was so much of the sun's reflected light on the vehicle that I couldn't see the stars. It was scary and eerie out in that dark abyss of space. Al said the most beautiful sight he saw on the flight was the view looking back at me. I was suspended outside the hatch with the full yellow moon completely framing me. The National Geographic did a painting of me against the full moon; it almost looks like a photo.

Al did a great job out there on the SIM Bay. When Houston had had a chance to digest the EVA, they got back to us. Dr. James R. Arnold reported that Al Worden had probably performed the first recorded repair of a scientific instrument in space. Earlier in that day, after he had experienced some problems with excess noise in the gamma-ray experiment, Al had gone on the EVA and did something—

nobody knows what. The gamma ray cleared up and worked beautifully after that.

That day had a salutary effect on everybody. The EVA had been successful; we got a report that the high voltage had been turned on to the subsatellite and that all systems were operational. Everything was better. As we were coming back to earth, the spacecraft was cooling off. It could have been that we were using less power. But the spacecraft was also just naturally cooling off. And we were adjusting to the environment. The daily routine was easier; sleep was better. I was feeling better all the time and really hoping that the flight would be longer, because it was a good simple life.

Just as we were really enjoying this life of traveling through space, it was rapidly coming to a close. So we had some mixed feelings. Sure, we wanted to see our loved ones and get back to earth. But this was such a great trip we were on. We had adjusted physically and psychologically. People often ask, "Didn't the spacecraft become smaller?" To the contrary, it seemed to become larger. On the way back there was never the feeling of claustrophobia. Of course, we got a little smaller. When we got back on the ship after splashdown, I had lost five pounds and the others had lost three pounds each.

The next morning, Houston roused us at about 6:03. The only bad news was that our TV camera on the lunar surface had cut out. It was panning and zooming in and out, getting views of the surrounding mountains; then Houston lost everything.

About 1:45, we got a query from Houston about the lunar eclipse we had been watching. At this point we were getting close to a total eclipse. We had seen the moon go through its phases. It had turned from a bright yellow to a dark orange as the shadow of the earth went across it. We saw a variation in color from a light gray to a burnt orange from one side of the moon to the other—almost like the old harvest moon. CapCom told us we were seeing the eclipse of the moon twice as big as anyone else has ever seen it.

The big public event scheduled for today was the press conference. Newsmen at the Manned Spacecraft Center in Houston who had covered the flight had submitted questions, and our little crew was going to knock down another first— a press conference in space. This was our introduction to the brand-new life that was going to begin in space and pick up momentum once we got ourselves back to earth.

The first question: "This last week we have shared scores of exciting moments with you. Which one would you most like to live again, and is there any moment which you would never like to repeat?"

Dave replied: "I guess the most impressive moment that I can remember is standing up on Hadley Mountain, Hadley Delta, and looking back at the plain, and seeing the LM and the rille and Mount Hadley and the whole big picture in one swoop . . . Al?"

"I guess I'd have to say, sort of, two events occurred," Al said, "which were exciting for different reasons. And I guess they were really the highlights of the flight for me. One was right after the LOI [Lunar Orbit Insertion] Burn, when we got our first look at the moon, and it was a fantastic, spectacular sight, and the other, I guess, was when the TEI [Trans-Earth Injection] burned so beautifully, and right after TEI; that was an awfully good feeling."

Well, being third in line could have been something of a problem, but I told the press that there had been a great many new thrills for me—the most impressive being the lift-off. It was then that I knew I was finally going into space, after years of waiting and training. The event that I would not like to repeat again was the time I fell down in front of the TV when we were deploying the Rover.

The next question had to do with the Genesis Rock. Dave described the way the white rock had been presented to us on a pedestal: ". . . if it is in fact the beginning of the moon, [it] will tell us an awful lot, and we'll leave it up to the experts to analyze it when we get back, to determine its origin."

Then, interestingly, Dave was asked whether the work load during the three lunar surface excursions was too demanding.

His answer, in short, was, "I don't think we ever reached anywhere near the limit of our physical endurance." He concluded that the hours we had spent on geology field trips and working in the space suits at the Cape had helped us with the EVAs. "I see no problem in the future with conducting three successive seven-hour EVAs. Neither one of us was particularly physically tired; I think fatigue is really in the mental regime in which you are concentrating very intensely for seven hours and you're pressing to do your best all the way through. . . . I think it is really more of a mental factor than a physical factor."

After explaining that the soft soil on the surface of the moon was five meters thick and not the expected one meter, "The mechanical task of doing the drill at that time seemed somewhat less important than seeking new finds in a new geological area," Dave said. "But in retrospect I think we have in fact brought back one of the most significant samples of the whole trip."

Al replied to questions about his solo orbits and his space

walk. When we were all asked about the problems, we all agreed that we had really had very few. The situation had been controlled so well it seemed like a tame simulation.

As I think back about all of this, it is surprising that all of us were presenting Apollo 15 as a flawless, serene exercise. Well, there would be one little problem that would come up later that would not be quite this easy to dismiss, but for the time being it was smooth as silk. And we all agreed that it had been worth the taxpayers' money.

I am sure there were some faint smiles in Houston when we told the press about the epic flight of Apollo 15. When you think about the staggering amount of support we got— the backup people constantly working with duplicates of all systems to solve problems that we had not discovered yet; the intricately organized information, the accessible expertise, the improvisation—no wonder we could fly.

Finally, we showed a TV picture of the moon just coming out of eclipse. In voice over the picture, we told them, "Houston, the moon is an orange ball now. Dull orange ball with a sort of gray area in the center, and on one side, opposing the side that's slowly coming into illumination, you will probably get more and more of the lunar surface exposed to sunlight. . . ."

Believe it or not, after we had stricken the agony of getting those core stems from the record, the subject came up one more time. CapCom asked us late that afternoon, "How long is the core stem at the present time? Did you break it down or is it still three sections long?" We replied that it was still three sections long and we had stowed the sections in the bag on the side of position 8A because it was convenient, and not, as Houston had suggested, in the sleep restraint that had the Command Module Pilot's pressure garment assembly in it.

FLYING HOME

"How did you get it in the 8A compartment, which according to our measurements from the ground is only thirty-six inches long and therefore apparently not long enough to hold the three lengths of core stem?"

This core-stem motif is instructive, if not diverting, because it tells you something about Houston. We may not have been the greatest explorers on the world's stage, but we had the greatest prompters. If you look at the Apollo flights—particularly 15, which was successful—from a certain perspective, you could say that the crews didn't have any problems. Houston had the problems. And nobody in the press queried Houston about that.

Having been challenged as to how we could get forty inches of stem in a thirty-six-inch bag, we had to say, "Well, it's sticking out a little bit, I guess we have to admit, but it's pretty well cinched down." With their masterly diplomacy, Houston said, in effect, to please put it where they had suggested if it wasn't too much trouble.

So we did.

Al shot a few stars that night in a mid-course navigation exercise, and then we went to bed. Splashdown was approximately 17 hours and 30 minutes away.

"Rise and shine. It's splashdown day." That was Joe. They didn't take any chance that we might not appreciate that it was splashdown day. Houston piped us the "Hawaiian War Chant"—a version done by Al Kealoha Perry—and the heart rates on the cardioscope jumped straight up in the air. Hawaii is where we are bound for, give or take a few hundred miles.

We woke up at 7:04 A.M., and at about 7:15 we got our first view of the earth. It was getting bigger and yet it was getting smaller. The crescent earth looked so much like the

105

moon that it seemed we were going in the wrong direction. There was a very thin sliver of a large round ball. The earth goes through the same phases as the moon. When we left the earth it was a full earth. When we came back it was a new earth; absolutely black like the dark of the moon, only it was the dark of the earth. You couldn't see it. It was just like the moon when we cruised into lunar orbit, a dark mass in the sky.

By 9 A.M. we are within 42,587 nautical miles of the earth and our velocity is up to 9,094 feet per second. We are looking forward to a mid-course burn of approximately 13 seconds. We don't need the SPS rocket engine—we will use the reaction control system attitude thrusters on the Service Module for the burn. It might be interesting to note that without the burn, according to Houston's calculations, we would continue into the entry corridor and could make it into the atmosphere on the trajectory that we have now. However, the mid-course will put us in the center of the corridor, and that is where Houston wants us to be. Without the correction, Apollo 15 would splashdown approximately sixty miles short of the landing point.

I am not at all tense about reentry, because Houston sees us coming and they know exactly what our angle is going to be. The only thing I'm concerned about is that we have the capability of rolling the spacecraft at the precise moment to ensure capture by the atmosphere.

The moment of the burn comes, and Apollo 15 is thrusting. It is a good burn of 21 seconds. Velocity is building and is soon 12,410 feet per second. We are now 2 hours 47 minutes away from the earth's atmosphere, 3 hours away from landing in the Pacific Ocean about 285 nautical miles north of Oahu. Our distance from the earth is 23,494 nautical miles. We get confirmation that we are in the center of the reentry corridor.

Soon two helicopters will be airborne from the *Okinawa.* They are *Photo* and *Relay* (their call letters). *Photo* will have the photographers and *Relay* will have backup swimmers and will serve as radio relay between the carrier and the spacecraft and the other helicopters in the area. There are also helicopters *Swim 1* and *Swim 2*, with *Swim* 2 assigned to deploy swimmers with the flotation collar that will be fastened around the Command Module to keep it stable in the water.

At about 1:57 P.M., I get the news that my mother and father have arrived in the viewing room at the Houston Flight Control Center. The brass is checking in with Dr. George M. Low, the Deputy Administrator of NASA, and Dale Myers, the Associate Administrator for Manned Space Flight, along with the Deputy Director of the Manned Spacecraft Center, Christopher C. Kraft, reported in the control room. Deke Slayton is with CapCom Bob Parker and the Apollo 15 backup crew at the CapCom consoles.

As we rush into the sensible atmosphere, which extends up to about 300,000 feet or 50 miles, for about four minutes you have this fireball effect. There is a total loss of communication. You look out the window and there is just this beautiful orange-yellow glow. The heat builds up to 5,000 degrees F on the outside of the spacecraft. It is charring the ablative material, and some of it is peeling off. You also have the glow of ionized particles that are streaming out behind the spacecraft. Dave and Al are monitoring the systems, they are in position to roll the spacecraft into the atmosphere manually if the computer doesn't do it. They have to control our entry.

We use the atmosphere to slow down; we are riding in and coasting as if uphill, against the atmosphere. I have the task of putting the camera in the window and turning it on so that we can get some good pictures of the fireball. We are coming in upside down and backwards—can you imagine that? I can see the glow out the window, but off to the right I can also see the Pacific Ocean. We are coming in over the Indian Ocean. I can see a few white clouds. We do that roll maneuver, I look out, and on the horizon are some snow-capped mountains. The first land mass I see on the earth as we come back are the snow-capped mountains of New Zealand. It's a beautiful sight. We're coming back from the mountains of the moon to the mountains of the earth.

It is kind of a physical endurance test, a traumatic experience, to go from zero G's to almost 7 G's during the entry period, during the fireball. For about four minutes you experience this. You have seven times your weight; you weigh over 1,000 pounds. It is physically impossible to lift an arm up to touch a switch, or move a circuit breaker. It was amazing to me that Dave was able to talk to Mission Control while we were coming in. I couldn't take a breath. I was living on the residual oxygen in my lungs. It felt like an elephant was standing on my chest. Couldn't move the diaphragm. Not painful, just a tremendous force on your body.

We were plastered against the couch. I felt I was on the verge of blacking out at 7 G's, but you could probably go to 10. We felt it so much because our bodies had become so lazy. The heart was lazy. We had lost muscle tissue; we had become flabby and weak. So, going from zero G to 7 G's felt like going from 1 to 14 on the earth as far as I was concerned. You start sensing G's almost immediately, and the buildup is fast.

When we hit about 60,000 feet the altimeter came off the peg and we started reading altitude. The atmosphere was slowing us down. The drogue chutes came out at 25,000 feet, and that really slowed us down, like a drag chute or speed brakes in an airplane. It doesn't jerk you, it brakes you—then you start oscillating wildly underneath the drogues. Then, just before you get to 10,000 feet, the drogues are released. You can see up there; you see them go, and you feel them go. You free-fall for a few seconds, wondering whether the main chutes are going to come out. At 10,000 the main chutes are out and they slow you almost completely.

We saw three chutes; then we saw that one had failed. It just collapsed. Well, Al saw it and I saw it, and at the same time the helicopter crews that we couldn't see told us we had lost a chute. The recovery crews had moved into position at 7,000 feet, and they were all around us, telling us, "No sweat, you can make it on two." We just about hit one chopper as it moved beneath us. All of a sudden the pilot looked up as we came out of a cloud right above him.

The chopper crews kept talking to us so much that Al and I could not communicate inside the spacecraft. "Apollo Fifteen, we are with you. Be advised that you only have two chutes instead of three." "WE HEAR YOU, WE HEAR YOU," we said. This was one of the busiest times in the flight. They couldn't help us at this point; it was most annoying, because they couldn't help us until we actually splashed down. Meanwhile, I realized that I did not and would not have time to stow my camera; I couldn't get to a pocket to put it in. So, I just took it off and put it between my legs. There we were with one collapsed chute, anticipating a hard impact, and I had this camera in a very sensitive spot. Also I had to have my fingers over the right side of the circuit breaker panel at the moment of impact. At splashdown I would push the circuit breakers to release the parachutes so that they won't pull us over and upend us.

We hit. It was very flat, a very positive impact. Al had been on the left couch flying us in, so he was in charge of this landing. Dave had flown Apollo 9 back in, and he wanted to give Al a chance at the command seat.

Al and I thought we had a rather smooth splashdown, but Dave thought it was twice as hard as Apollo 9's impact. I released the chutes the instant we hit. Actually, we should not have released the chutes. There wasn't any wind blowing, so the chutes would have just come down by themselves and lain there on the water. The frogmen were able to recover only two, and they did not get the chute that collapsed, so NASA never learned what caused it to fail.

It's so good to be back to the good earth. We are sitting there eagerly anticipating the frogmen who are on the way to open the hatch. We are looking for them to appear in the window. Suddenly, Freddy the friendly frogman who has trained with us in the Gulf knocks on the window. We have been there in our underwear, and now we have our clean cover garments on—we have not worn them since we left. We feel so relaxed, so relieved that now we are back on earth. We stay there in the Command Module like forever. It is bouncing a little bit.

Finally, they get the hatch open and the life jackets in there, and we put them on. Dave goes first; I go next; Al goes last. I get into the raft and, man, it's great. Beautiful. Nice and warm. The first thing I do is dip my hand in the ocean and put the water on my face. Just to feel that water—water from the earth on your face—and to breathe that air.

4

REUNION

Back to the blessed earth and down in the warm, lovely, calm sea. Finally we were all on the raft. Lying back, breathing that good sea air, we felt relaxed, but there wasn't much time because the chopper moved in. It hovered overhead about fifty feet up, beating us with downwash, and then took us up in the rig that they call a Billy Pew Net, one at a time. The whole thing was just going too fast—I wanted to slow it down, do it in slow motion so I could really savor it.

I was hoisted up to the chopper, and there was Dr. Clarence Jernigan, the NASA flight surgeon. He'd checked us out before we left, and here he was waiting for us. They gave us clean blue NASA flying suits, dry, clean tennis shoes, and Navy flight caps with the scrambled eggs on the bill. The chopper kind of circled out there while we were hopping around on one leg, putting the flying suits on over our scroungy long underwear. We had been in space for twelve days, and we were not very steady.

My primary concern was: What am I going to say? I hadn't written a speech, and I knew that when we got to the carrier I had to say something. During the flight, Dave had the left couch, and I had the right couch, so we had agreed to make our remarks in that order. Sometimes it was difficult, because I could have my speech ready, and then Dave would use my lines and I would have to improvise. We were kidding each other about what we were going to say and suddenly we were landing on the carrier.

We wanted to look military, not scraggly; Dave's big complaint about all the previous flights was that the crews never saluted in unison. We had gone over this many times. Dave had said, "I want you all to salute when I salute." We said fine, that's the way it should be. So when Dave started going out the hatch, I said, "Okay, Dave, we are right behind you. Wait until we get out there to salute." But he got out on deck and saluted instantly; then Al saluted; finally I saluted, quite a bit later.

Dr. Robert E. Gilruth, Director of the Manned Spacecraft Center, was there. He had never been out to a recovery before. Dr. Gilruth had been a tremendous force in getting the Manned Spacecraft program going; it was great that he could be there. Gen. Pete Everest was there, the man who broke the sound barrier. He has always been a hero of mine. There were a lot of congressmen there, including Olin Teague from Texas, who was as close to being my congressman as anybody I can think of. We went down the line greeting all these people and all the Air Force and Navy brass. It was probably the warmest receiving line I had ever gone through. They had big smiles on their faces; they were really happy to see us, and we were really happy to see them.

About this time I started to feel that I didn't have the balance that I should have. It wasn't just that the ship had some motion; I wasn't as steady as I should be. Then I realized that I hadn't cleared my ears coming down. I was so busy, and the fact that we lost a chute had made us come down much faster. So there I was in this receiving line with my ears plugged up and uncertain about my balance.

My speech was very simple. I told them how thankful

we were that we had been able to make the trip and how grateful we were to all those people back here on earth who helped us. I mentioned my Navy training and said that I had been on ships before but I had never been so happy to be on a ship as I was now.

It was an astronaut-type speech. I didn't say anything about the religious experience I had had on the moon. I still hadn't gotten this together in my mind. At first I just talked about the technological voyage, not the spiritual voyage, not how I had been changed in my heart and in my spirit.

After the speeches we went into the wardroom, which was our headquarters as long as we were aboard ship. The doctors were there, and they asked us what we wanted for lunch. We all agreed that steak would be fine. It was good to be able to bite into something solid after all that gruel. But, really, food didn't interest me as much as I thought it would. Dave and Al had been talking about ice cream all the way back. This is not what I dreamed of on the way home.

It was so good to get back on earth. To be able to move around under a 1 G environment was such a pleasant sensation. I was hoping that we'd have a chance to take a shower, but we had to wait a long time for that: the doctors wanted us to complete the splashdown physical first. They didn't want us to wash the moon dust off. They took blood from us right in the wardroom before we sat down to lunch.

Lunch was great. Our systems didn't have any trouble at all getting reaccustomed to good old earth food. But I don't think we ever did get that ice cream that Al and Dave had been dreaming about. After we ate lunch, we were invited down to the hangar deck. There was a little ceremony there and we had the chance to speak again, which seemed kind of ridiculous because we had already spoken on the flight deck. However, I guess they wanted all the rest of the guys on the carrier to see us. The ship's chef had baked a tremendous cake; it must have been eight feet square. It said APOLLO 15, WELCOME HOME! We cut it and had a piece of cake with them.

Then we spent the rest of the afternoon going through physical examinations. It must have taken four hours. First, using cotton balls, they swabbed us on the bottom of the feet, in the armpits—any place that microorganisms might be growing. Then there were many medical tests. One of the most demanding was the ergometer: we got on a bicycle and began pumping against increased resistance, and they monitored our heart rate and the efficiency with which we were breathing oxygen. Before the flight I think I could go for about eighteen minutes. But this time I might have gotten to fourteen or fifteen minutes. The heart just wasn't performing; it was very lazy. In fact I felt rather dizzy at one point—the doctors thought I had a lapse of concentration or memory, or something like that.

Then they put us through the LBNP (Lower Body Negative Pressure) test, which involves putting you into a box that seals around your middle. You are lying down horizontal and they evacuate the air to successively lower levels. Then they study the way your heart responds. The LBNP test confirmed what the ergometer had shown earlier, that we were in pretty bad shape. In fact, I think I just about passed out. The room started to whirl around. Both Dave and Al tested about the way I did. After the tests, I finally got that coveted shower—one of the longest showers I have ever taken in my life. Man, finally. That was the best part of getting back home: to strip off all those dirty clothes and feel you smell reasonably good again. You know, I guess I wasn't as cruddy as I thought I was going to be. A lot of that black moon dust just seemed to have cleared off my body into the atmosphere. But I stayed in the shower as long as I could, right up to the time I was scheduled to go to dinner with all the brass.

We turned in early, about 8:30 P.M. That was the first of many strange nights. I woke up several times to go to the head, more than I normally have to, and it was really peculiar. The first time I woke up, I just felt weird, as if my body were inclining rather than flat. My head seemed about 30 degrees down. It was pitch black in the stateroom; I sat up and got out of bed and managed to stagger in the darkness through those Navy hatches into the head. When I got back in bed, I was tired enough to doze off right away. Then, two or three hours later, I would wake up again with this same sensation and go back to the bathroom.

That wasn't a good night's sleep. For some reason, we were right under the hangar deck, and at about 5 A.M. they started clanking those chains. That was a horrible noise, like being in a tin can that somebody is beating on; none of us could sleep. I wasn't disoriented, I just had a different orientation from any I had ever experienced. I wasn't worried about myself, but I was kind of surprised.

We had been steaming toward Pearl Harbor all night, and the next morning when we got up on deck we were there. All we had to do was take a chopper over to Hickam Field, where they had another welcoming ceremony ready to go. We landed right on the field, and there everybody was. As I remember, Carl Albert, the Speaker of the House, was there to greet us, along with some congressmen from Hawaii. There must have been two or three thousand people standing in front of a great big platform for radio and TV.

Incidentally, that morning I mentioned my vertigo to Dr. Jernigan. He said it was strange, but he didn't seem to be inclined to pursue it.

I felt mildly tight and a little off balance in front of my Hickam Field audience. I told them we had been in Hawaii many times and that it was most appropriate to come back from the mountains of the moon to the beautiful mountains of Hawaii. I said "Aloha," and then I told them that when we made the TLI Burn and started on the way to the moon I had been able to see all the islands of Hawaii in my window. I said that was the most beautiful sight I had as I left earth. They were very responsive. We finished our remarks; then we got aboard a C-141 on the other side of the field and were off, bound for Houston.

They had a cramped little compartment in the C-141 for us—it was like traveling in a closet. It was even smaller than the Command Module. We were getting pretty sick of each other's society by now, but fortunately we had a task on hand. Six hundred and fifty envelopes were among the personal items we'd taken to the moon with us. They had been postmarked before we launched and postmarked again after we boarded the carrier *Okinawa*. Dave had made the necessary arrangements. Each of us had to autograph every envelope, so we were busy for several hours.

It was after dark when we got over Houston. A storm had just passed Ellington Field, and there was a lot of water on the runway. It had occurred to me that it would be ironic to rack it up landing in Houston. Before our flight they had always separated Apollo crews because they didn't want to wipe out a whole crew at a time. Dave and Al were still talking about ice cream. There was some question in my mind as to just who might be in Houston. I was hoping that my folks had stayed there during the flight. It would have given them a chance to visit and help Mary with the children.

We arrived just after the storm and taxied up. Deke Slayton, our big boss, was the first one in the airplane. "Welcome back, guys, you did a great job." We left the airplane together, and our loved ones ran up and surrounded us. Mary and I kissed and had a warm embrace. I think I said, "Hello, Schnook, I'm back." I can't remember the family ever looking any happier. I had Joy and Jimmy in one arm and Jill and Jan in the other. Then Mom and Dad rushed up to us; I was pleased they were there.

As I climbed up on the wooden platform I thought this was one speech I was really turned on to. It was great to see all our friends out there in front, to look into their faces again and see the smiles. Here were the people who always come out when the crew returns, no matter what hour of the day or night—the folks of Houston, particularly those who work at NASA. I was thinking of all those guys, how they supported us when we were on the flight, how hard they were working on the ground. When you have a cast of thousands that you are dependent on, you cannot help but have a constant feeling of gratitude. I had been on the other end—you know, coming out to welcome the crews. It means a great deal to see everyone coming out. I know.

They had worked a lot harder than we did, really, and probably got a lot less sleep. Of course, after splashdown, the work was essentially over for our crew, and the guys could go back to a normal routine. So we necessarily kept our remarks brief so we could all get to our homes. I mean all of a sudden, our home is not the moon, it is the earth. NASA has assigned a protocol and PR expert to each astronaut family to protect them from the pressure of the press and from other outside interests. Our man was Bill Der Bing, and he had a station wagon waiting for us.

It was just the greatest reunion with the family. The children were almost beyond words. "Daddy, it is so good to have you back." Mother and Dad stopped by the house for a short while, but they knew I wanted to relax with Mary, so they went back to an apartment that some friend had loaned them.

I was a little hungry, and Mary asked me if I wanted something to eat. "I'll have some soup," I said. So she broke out some of Chalet Suzanne's soup, lobster bisque, to keep me from being homesick for the moon. It was kind of appropriate to come back and have a little Chalet Suzanne soup.

There were some of the press there in the house, waiting to take pictures. They apologized for intruding and reassured us that they would take a few pictures and then be gone. They took a shot of me sitting down and eating soup. The kids were tired, so they went to bed. Then Mary and I turned in too. This was one of the things I had dreamed about late in the flight. I was eager to get back to Mary. She said, "I've missed you." And I said, "Yes, Schnook, it's good to be back on earth."

I went to sleep very easily that night. Per normal, I woke up early, about 6:30. My folks were leaving that morning, and I wanted to see them before they left. I didn't get any breakfast, because I was to report without breakfast for a physical—blood test, urinalysis, electrocardiogram—before we started debriefing. Many times I went back into the LBNP box over the next days so they could check the response of my heart. After three days I lost the feeling that my body was tilted at an angle, but I was still having a problem with this lack of balance. It lasted ten days, but before they ever got around to diagnosing it I had completely recovered.

Most of our time during these next days was spent in debriefing. We used the flight plan as our schedule, and we were debriefed individually on our specific assignments. I covered the Lunar Module, particularly systems problems. We just saw one side of the mission—in many instances the people on the ground knew more about the problems than we did. We worked the flight plan from Launch right through Splashdown. Occasionally, we'd break and go to the doctors for a special physical.

We had a big press conference coming up five days after splashdown. Since we were the first crew that was not required to go through quarantine, we actually met the press before debriefings were completed. Another thing, we had been offered the opportunity of selling our personal stories, our immediate first impressions of the trip, to *The New York Times* for $1,000 apiece.

They talked to the three of us separately and picked our brains about our first impressions of the flight. I thought it was an imposition. Here we were, not back to normal physically and still involved in debriefing the flight. Yet we had to go over to the motel and spend an hour or two with these *Times* people. I got into a little trouble about this. I told them exactly what I felt.

I told them how I had been affected by the beauty of the mountains of the moon, and for the first time I told how I had felt the presence of God. Then I mentioned my vertigo. I told them that when I woke up I had had the feeling that my head was inclined 30 degrees, and that this had lasted about three or four nights. But I admitted that I still had the unbalance. When they asked me how I felt generally, I said, "I don't feel normal." I told them, "Something is not right yet."

The *Times* wrote up the interview, and it went through NASA PR for clearance. Since NASA had not released any of this information, their PR people decided it had to come out of the article. The *Times* people were a little chagrined that it was handled this way.

Dave was sensitive at this point because the medics kept telling us that we were not back to normal. After they took the blood samples, they found that our potassium levels were down. As soon as Mary heard about this, she went right down to the drugstore and got some potassium pills and saw to it that I took them.

Three or four days after we got back, I got up early one morning and went over to the gym and ran around the track. (On earth, I usually try to jog at least a mile a day.) I didn't feel as if I were 1 G while I was running. I felt that I was ½ G, and I had the sensation of floating. It was the strangest feeling to float along with my heart doing very little work. I jogged for the next few days, too. I don't think it helped.

The next weekend, on Sunday, I went off and played tennis, and toward the end of the first set this feeling of un-balance went away. All of a sudden, my balance was back. The next morning I requested that they run a test on me on the ergometer and check me out with the LBNP. They did, and I was back to normal.

I don't know how much permanent heart damage I suffered from the little attack I had after I left the moon, or how it related to a heart attack I had later. They already had pretty positive evidence that the loss of potassium could have been a contributing factor. On flights since then, 16 and 17, they put an overabundant supply of potassium in the food. This solved the heart problem, but evidently it created a lot of problems in the intestinal tract.

We finally completed our debriefing. Of course we were aware that it was purely technical rather than human or psychological. Very naturally, we had withheld our individual reactions to the flight. This is not what NASA is particularly interested in, and the crews have always held back their deeply personal feelings for their own books and magazine articles. At this juncture we were face to face with considerations of the ultimate financial remunerations of an astronaut. What is all this worth to Time, Inc., Field Enterprises, or whatever? Is being an astronaut worth half a million dollars? One million? It surely is worth a great deal in convertible PR value, but it is going to take me a long time to figure this out. It just depends on how you want to use it, how you want to convert it.

All this time, as I reflected on what happened to me, I was trying to put together the message I wanted to share with the world. My first experience in witnessing had come as a result of a request from the Nassau Bay Baptist Church, our church in Houston. (About three or four months before the flight, I had decided to join this church.) They didn't put any pressure on me or the children. Brother William H. Rittenhouse, a great friend and a tremendous pastor, just asked me one day about six months before the flight if I wanted to share a word of what Jesus meant to me with another church on the west side of town, the Willow Meadows Church.

I told Bill I didn't know; it was something I had never done before but I would think about it. I finally agreed to go over to Willow Meadows and speak. I told them my story —about how I had accepted Jesus Christ as a youngster, and about my accident, and about how much prayer had meant to me and how much God and Jesus Christ meant in my life. Bill Rittenhouse was impressed; he wanted me to share it with his congregation. So I spoke for about ten minutes during a Sunday morning service at my own church.

All this time I was developing a close relationship with Bill Rittenhouse. He was interested in our family, and he had a tremendous personal testimony of his own, so the children and I had been going there every Sunday. I was satisfied that they had a good Sunday-school department too. I used to go with the children and monitor the classes to see if they were getting a good program.

Occasionally, Mary would go with us. Mary had always belonged to the Seventh-Day Adventist Church; she was even teaching a high-school group there. Sometimes she would take the children to her church, but I did not want to see them confused, torn between two churches and having to decide which parent they were going to please.

A combination of factors had led to my decision to become a Southern Baptist and join the Nassau Bay church. One, of course, was Bill Rittenhouse. I could go to his church and enjoy the service, and I was impressed with the people. They were interested not only in the church but in the community, trying to reach out and share what they felt about Jesus Christ with others. They had a good program for children. They were getting something out of church; they were learning something. And there was an unexpected influence out of the past.

In the spring of 1971, I had taken the family with me on a trip to Florida. We drove through the little town of New Port Richey, where I had lived years ago, and we saw the little church where, as a youngster, I had stepped forward to accept Christ. We didn't accidentally find it; we sought it out to see if it was still there. I wanted to see what kind of church it was, because I thought it was a Community church or a Congregational church. And here it turned out to be a Southern Baptist church. I had really forgotten that.

Then there was the influence of two other fellows in the astronaut program. Bill Pogue and Jack Lousma, two astronauts who are close friends of mine, always seemed to represent my ideal of Christians. I had an experience with them in Hawaii on a field trip that I have never forgotten. We had climbed Mauna Loa, and I came back that night with some sort of bug. I was really sick. Well, Jack and Bill were concerned about me. The two of them took me over to an Army camp that was close by, and they found a doctor there who looked at me and gave me some pills. Then they helped me get back to the cabin. I appreciated their looking after me, and I knew they were Southern Baptists, actually members of the Nassau Bay Baptist church.

All these things sort of fitted together in my mind and had something to do with my becoming interested in the Southern Baptists. I wanted to be a member of a church before I went off on this mission to the moon. The whole time this was working inside of me, I was encouraging the children to make a decision. They seemed to be happy in the Sunday school, getting a lot from it, and I felt it was time for as many of the members of the family as possible to get together in one church.

One Sunday morning I told the children that I was going to make a decision for Christ and asked if any of them wanted to go with me. Jill, my second daughter, was the only one. At the end of the morning service, when Brother Rittenhouse invited us to make a decision for Christ or to rededicate our lives, Jill and I stepped forward. Jill accepted Jesus Christ as her personal Lord and Savior, and I rededicated my life to Christ. We both decided that we wouldn't be baptized until I got back from the flight.

I feel now that the power of God was working in me the whole time I was on the flight. I felt His presence on the moon in the most immediate and overwhelming way. There I was, a test pilot, a nuts-and-bolts type who had gotten rather skeptical about God, and suddenly I was asking God to solve my problems on the moon. I was relying on God rather than on Houston. Then there was my powerful desire to have a service on the moon, to witness. All this time God was taking over my life, and I didn't even realize it.

The Lord has found me a stubborn, hardheaded man. Mary could see that something was happening inside me, but I didn't realize it. She has told me that she asked the Lord not to give me a minute's rest until I had completely surrendered my life to Him. And, really, I didn't have any peace. Neither of us did.

At this point, back in Iuston with the great flight to the moon over, and before I had realized that the highest flight of my life was still ahead of me, I began to reconsider everything about my life. I thought back to that evening in New Port Richey, Florida, when I had accepted Christ at the little Baptist church. I tried to retrace my life from the beginning. For the first time, I was looking at myself introspectively to find out who I was and what was happening to me.

5

EARLY YEARS

I was born in McGee Hospital in Pittsburgh, Pennsylvania, on March 17, 1930. My mother and father lived in Beechview in the South Hills section of Pittsburgh. My father was a steamfitter in the Carnegie Museum, and some of my earliest memories are of waiting for Dad in this tremendous place. If he wasn't ready, I would tour the museum, looking at dinosaurs and other prehistoric creatures and every other strange and unbelievable thing you could imagine. The basement of the building, where Dad ran the power plant, must have been half a mile long. Many times when I was visiting him, I remember watching the great steam turbines generating power for the huge building.

From my youngest years I remember that Dad was tired of this job and wanted to do something else. He hated the terrible winters, and we all had horrible colds, and I know he dreamed of a change of scene—maybe going to Florida to live. Dad always had very ordinary jobs. He was never happy for very long with any of them. He never made much money, but somehow he was able to afford nice homes. Dad may have been a blue-collar worker, but the people who lived around us were white-collar workers.

Dad is a natural householder and homebody. He was the one in the family with a strong desire to have a nice place, and he was very particular—the house had to be painted; he was always looking for termites. At seventy-seven, he still puts himself out, still mowing the lawn, and still looking for termites. This may be a carryover from his old Irish background. Speaking of that, Dad always drank—but I've almost never seen it affect him. I said he was a homebody; he liked to drink at home.

Curiously, Dad was a much more military type than I am. I think he would have loved the Air Force, and I wish he had made it into aviation in World War I. He would have done very well. My Dad may well be smarter than I am. I was always thinking about how I could get out of the service, but Dad remembered the service nostalgically. He volunteered and served as an infantryman in World War I. As a young boy, I remember looking at the medals that he won. He was very proud of them, and he carried them with him wherever we moved.

Dad isn't as big as I am—he's a little man—but he's wiry, with muscles like steel, really solidly built. He was a fast runner, and he won many medals ice-skating when he was young. When he was teaching us things he was very direct. When he wanted to teach me to dive, he took me down to the Allegheny, grabbed me, and dove with me into the river—down about ten feet. I was sure I was going to drown. I had never been that deep before, and I can remember looking up and seeing the yellow murky water and wondering if I was ever going to get to the surface. He just wanted to expose me to old river-type swimming. I don't believe he was as affectionate with us as I am with my son. He's not the type to put his arm around you very often. My mother was the affectionate one.

The South Hills section, where we lived, is a very hilly area of Pennsylvania, where the Monongahela and the

Allegheny rivers come together, weaving through the hills and deep valleys. It is very exciting country, and the complexity of the land features always makes me feel I would like to fly over and explore from above.

I got interested in airplanes very early. A neighbor of ours in Beechview gave me a beautiful model airplane that became one of my treasured possessions and started my interest in building model planes. Then, in second grade, we built an airplane that was large enough for several of us to stand up in, about four feet high and ten feet long. Come to think of it, it was about the size of a Command Module. We were very proud of it, and all the kids had their pictures taken by it.

When I got a little older, we always went for a Sunday afternoon drive, and frequently we'd go out to the county airport to watch the planes take off and land. This was near some property that Dad owned and dreamed of building on someday. My early interest in airplanes and hearing Dad's stories about the service must have prepared me to take the direction that I finally did.

Chuck came along when I was four. I don't recall any difference in temperament between me and Chuck. He was basically quiet and gentle—about the same personality that I have. We used to fight all the time, like kids will, and pound on each other, particularly on trips. I'm sure that we made a lot of vacations absolutely miserable.

While we were living in Beachview we went to the Lutheran church, where a Reverend Burkie was the pastor. I always had a few lines in the Christmas plays, and when guests came to see us Mother always asked me to tell the Christmas story. When I was a little boy this was my mission, and I took great pleasure in performing for them and getting their applause. You might say it was a very early pulpit.

Just as I entered school, we moved to Brookline in another section of South Hills. We lived right across from the DePaul Institute for deaf and dumb children, and nearby was an orphanage for boys. They had great facilities: swings, tennis courts, seesaws, a ball park, everything. I guess Mother had met some of the sisters, because frequently my brother and I would be invited over to swim in their swimming pool. The orphanage also had a farm project for the boys, so we would tour through the barns to see the pigs and the cows. When our friends came to visit us, we'd always take them over to the orphan home. There was a tremendous acreage of woods around these institutions, so we had everything—a place to walk and hike and make forts, and trees to swing on.

About this time I went into the first grade. It was very traumatic for me. I didn't want to go to school. It got to the point where I even dropped out for a while. I hated to be pulled away from the security of my home. My friends were in the neighborhood and at the orphanage, and mixing with all those other strange boys and girls was not fun for me. At this point in my life, I wasn't good at any subjects.

I was a "family boy" then, and I guess I always have been. I haven't said much about Mother, but Mother and I have always been very close. She called me James. I wrote a letter home to my mother and father every day that I was in the Naval Academy—and Mother still has every one of them. The whole time I was growing up, Mother was always interested in my contributing to the family welfare because Dad didn't make much money. I worked until I went away to the Naval Academy, and in the early days I always gave the family half of everything I earned.

Mother had a knack for making contacts. Just as she knew the nuns who let us go swimming, she managed to meet a man in the Beechview area who imported coconut direct from the Philippines, carefully packed in containers. I'd walk a mile over to his place, pulling my little red wagon, load it up, and walk back. Then I'd sell the coconut from door to door. This was my most interesting early job. I also sold magazines, peddling them on the corners and from house to house. *Look* was one of the magazines I sold, but it was never very profitable and I gave it up.

I always thought of Mother as being a very busy woman. She was always doing something. It wasn't that she was cleaning up the house all the time, though it was always neat. She was always involved in all sorts of projects. She created a friendly atmosphere; we always looked forward to coming home from school. I could tell as soon as I came into the house whether Mom was home or not; and she was usually there.

Mom was intuitive and inquisitive about everything. She was about four years younger than Dad, but she seemed much younger. Dad was more set in his ways and she was more flexible. When she tucked us in at night, she used to kneel beside our beds and say our prayers with us. I sometimes wonder if our kids will see that much difference between Mary and me. Actually there is eight years' difference in our ages.

Looking back on it, I know we had a good life in Pittsburgh. It was one of the calmer, nicer periods. We had relatives all around us, and we lived in one area for a long period of time. Both Dad's mother and father were dead, but he had a brother, Alex, and a sister, Anne (who lives in Atlanta now). Mother had five sisters and one stepbrother in Pittsburgh, so I had all these aunts and uncles. And we had cousins of all ages and descriptions. Mother's father and stepmother were alive. They had one of those typical old Pittsburgh houses up on a steep hill, a big brick house with a wide porch and a swing that was great in the summertime. There were lilac bushes around, and old-fashioned furniture inside, and a big basement where they had their homemade wine stashed away.

I remember all the family gathering in the house, with the old folks playing pinochle and my cousins and myself either out swinging on the porch or down in the basement or traipsing around the block until we got so tired we'd just pass out. Dad loved to travel and he loved to go on vacations, and frequently the whole family would go off for a couple of weeks together. Maybe there would be fifteen or twenty of us. We'd rent a big cabin where we could all be together, maybe at Yankee Lake or Oak Orchard, just to the west of Rehoboth Beach, Delaware. We would spend the days swimming and fishing and crabbing and have some great times together. I often regret that we moved away and left all the relatives and that great family feeling that we had.

To get back from Oak Orchard we had to cross Chesapeake Bay, and we used to take the ferryboat that came into Annapolis. On that ferryboat ride I got my first view of the Naval Academy. Dad even said maybe I would be interested in going there some day. Isn't it strange how the pattern is formed, how the most trivial and accidental events set up fateful decisions that shape our lives?

One summer we decided that we would strike out for Florida. We drove down the peninsula through Delaware and Virginia and crossed by ferry at Cape Charles. As we got farther south, it became so unbearably hot that with one accord we turned right around in the middle of the road and came back to Pennsylvania. The next year we tried it again, but this time we used the inland route, and I remember that I had the job of being navigator. We made it. We visited St. Augustine on the way down, and we all went out to the ancient fort and to Ponce de Leon's Fountain of Youth. I had a little of that water.

Florida became an annual thing with us; the more colds we had, the worse the winters were in Pittsburgh, and the more exhausted Dad got with his job, the better Florida looked. We traveled down the West Coast and spent some time at Clearwater, which we loved. Looking back, I know that Dad was more and more tempted to leave his job. But there were a lot of ties in Pittsburgh, and my brother and I were pretty happy there.

During the winters we had quite a time commuting from Brookline to the Lutheran church in Beachview. Several times one winter we walked the five miles through deep snow. It was quite an expedition, but it was fun for a young boy to walk five miles to church and five miles back—and it seems to me that we got special recognition when we accomplished this.

During these years, Dad was quite a disciplinarian. If I crossed him I knew that I was in for a rough time. I remember that one time he and I were down in the basement of our home in Brookline, and he thought I had done something that really outraged him. He lashed out and actually hit me across the face. This was one of the few times he ever struck me.

Dad was really funny about my social life. The atmosphere at home did not encourage dating, and Dad set out to instill fear of the opposite sex in us when we were very young. "All these girls are out for your money," he'd tell us. "They are gold-diggers. And they'll give you infectious diseases." The truth of the matter is that I didn't have any gold; any money that I earned I split fifty-fifty with the family until I left home.

Far from being in any danger of catching infectious diseases, I was afraid to go out with a girl. I can remember one time in Pittsburgh when I was invited to a girl's birthday party. I was probably eight or nine years old. When I got there they were playing spin the bottle. The idea of having to go in the back room and kiss a girl was revolting, and I left the party pretty early. It is rather amusing now to look back on the things my Dad told me about girls.

I'm not saying he was wrong. If I had become infatuated with a girl, I could have wasted a lot of time, decreased the amount of studying that I did, and I might not have gotten where I am now. As it was, I really started studying when I was in the sixth and seventh grades.

In August of 1941, when I was eleven, Dad finally got the family out of Pittsburgh and down to Florida. Deep snow, one last batch of colds, and one more winter in the boiler room did it. Dad packed us into the Terraplane or Hudson or whatever it was (he was always getting offbeat cars), and we drove down to New Port Richey. We looked all around and finally decided to buy the mayor's house. It was modest but very beautiful, with screened-in porches front and back and a big fireplace. However, we had one serious problem— Dad couldn't find work.

Can you believe it? There we were in Florida, and Dad couldn't find a job anywhere in New Port Richey. He was forced to go back to Pittsburgh. This decision involved a tremendous amount of anxiety and a lot of tears in Mother's eyes as she tried to understand why things were working out this way. Well, Dad packed up and we settled into our new home, and I took over the role of the man in the house. It was a very maturing experience for me.

The school people weren't certain which grade I should be in, the sixth or the seventh, because I had originally started school at midyear in the Pennsylvania school system. So I spent half a day in the sixth grade and half in the seventh. In the mornings I was in the grammar school and in the afternoons I rode my bike to the other side of town to the junior high. I had bought the bike from the mayor's son. It was the first two-wheeler I had ever had, and I was proud of it.

Chuck and I used to pedal out four miles to a fish house on the river and buy fish for the family dinner. On one Good Friday we went out early in the morning. I was pedaling and Chuck was riding on the handlebars. We had bought a fish and were on the way back when he dropped it. When he tried to slip off, his foot caught in the spokes and he went right down on his face. He must have had his tongue out, because when he got up his tongue was split right down the middle and he was spitting blood all over the road. I finally got Chuck and the fish back home, and fortunately there was a doctor nearby. I remember Mom holding his tongue out while the doctor stitched it back together. We had planned a picnic for that Friday, but Chuck felt so bad we stayed at home.

The principal of the school where I went to the sixth grade was also a Methodist minister. He was the one minister in town that we knew, so consequently the family ended up going to the Methodist church. Before I went to Sunday school each week, I would go by and visit the pastor and his wife. We became good friends, and they developed my interest in young people's work. So we were settled into New Port Richey, and Dad was back in Pittsburgh fighting another winter.

One night Mother and Chuck and I were walking through this little town, trying to get the evening breeze. We just happened to pass this little Baptist church and noticed a revival sign out front. I don't know what drew us in that night, because we were going to the Methodist church at the time, but something prompted us to attend the service. At the end of the meeting, they gave us an opportunity to make a decision for Christ. They called it an invitation, which was a strange word to me at that time. Something just touched my soul, causing me to get up out of my seat and go forward with some others. I noticed that one of the young girls who moved forward with me that night was in my class at school.

I had tears in my eyes. It was a happy moment when I could give my life to Jesus Christ and know that He was my personal Lord and Savior. This began a relationship that has continued every day of my life since then. I have not been steadfast over the years. I have wandered, and at times I became skeptical about the power of God, but the Lord has always brought me back to that early experience in my life, the night I came to know Jesus Christ. It is still a mystery to me how it all came about, but I know it might not have happened if my mother and father had not been Christians— if they had not had the patience to guide me and the loving concern to lead me down the right path until I could make the decision myself.

You know, we never went back to that little church again; we continued to go to the Methodist church. I became very active, particularly in the young people's program on Sunday night. I missed my dad, of course, but people were friendly in that town and it was really a very pleasant life. I even took voice lessons. A neighbor three or four houses down taught voice, and Mom thought, since I hadn't been able to learn to play the piano back in Pittsburgh, that maybe I could learn to sing.

World War II interrupted my vocal career. On December 7, 1941, the Japanese attacked Pearl Harbor. It startled us in our little haven there in New Port Richey. But soon things started happening. The Air Corps began building bases all over Florida, and before long Dad had an opportunity to come down and work at the Orlando Air Base. For the rest of the school year, Dad worked in Orlando all week and commuted to New Port Richey by bus on weekends. He took my bicycle to Orlando with him. It was the only transportation he had to get to and from work.

In May, when school was out, the family moved to Orlando to join Dad. We lived in Colonial Town, about two miles from my school. I remember that several fellows who were great athletes at the school lived right in the neighborhood, and they became my friends. I don't know why, exactly, because I was supposed to be a brain at that point. Anyway, there were Harry Hardaway and Hickson Cheeks and a couple of other guys who became real heroes to me.

They knew the girls in the neighborhood, so I got to know them too. A couple of beautiful gals lived down the street. One of them had her own car, so she'd invite us over to have coffee in the morning and then drive us to school. What a great setup, just the most perfect arrangement you could ask for, to have a ride to school every morning with those good-looking gals.

Just before school began, I was able to get a real high-class job. I worked weekends for the Frankels, a Jewish family who ran the Orlando Bargain House. It was located in the black district, so of course most of our customers were black. We sold everything—shirts, hats, shoes, underwear, clothing—at bargain prices to people with low income. It was really an education to work as a salesman with this sort of clientele. We had good relations, and I enjoyed it. I remember going to work at 8 A.M. Saturday morning and working until 9 or 10 Saturday night. I'd eat my lunch in the back room—a bottle of pop and some sandwiches Mom had made for me.

I got tremendous encouragement in science in the junior high school, and I'm sure this gave me the confidence and the interest that was essential to my career later on. I had a great science teacher, and she gave me the highest grade she had ever given on a final exam—100 percent. The other students received their papers first; then she made a little speech and gave me mine. I never will forget that day.

My history teacher in junior high, an elderly lady I was very fond of, also taught Sunday school at the Presbyterian church. I started visiting and then became a member of her Bible School class. Soon Chuck was there too. We both joined the Presbyterian church, and I remained a member of this church for a number of years. However, we weren't destined to stay in Orlando much longer.

About this time my folks decided it was just too hot in Florida. Dad was getting itchy feet again—maybe there was some dissatisfaction with the situation at work. Anyway, he was able to secure a transfer to another government facility in Roseburg, Oregon.

We packed up everything in September of 1944 and moved west. When I say "everything," it wasn't much, because we sold most of our belongings before we left Orlando. We still didn't have a car, so we traveled by train. Early one morning, Dad woke me up, saying, "Look, look out on the horizon." I could see the Rocky Mountains.

In Roseburg we moved into a little motel building down by the river, all four of us living in one room, and Dad went to work as an attendant in a mental ward out at the VA hospital. Sometimes he would come back at night with funny stories about his patients. Other nights he would be so depressed and exhausted that he would sit there silently, and Mother, Chuck, and I would try to cheer him up. He had to feed some of his patients. Some he had to take to the bathroom, and some would have to be tied up so they wouldn't injure themselves or others. This job was not in Dad's line. He gave it up and found another as janitor in the local school where I was a student.

This was sort of weird, but I enjoyed having him there. I can recall going back to the school many times in the afternoons and in the evenings to help him out. I'd help him mop the gym, or we'd work together cleaning the bathrooms. Speaking of the gym, one of my most vivid memories is of a gym teacher in that school who was an ex-military man. He insisted that we run at least a mile each week. It was a real chore for me, and I hated it, but for the first time in my life I was in an exercise routine. Then I managed to break a toe wrestling—which is extremely difficult to do. This saved me from the running for a while.

We were still just camping out in this little motel room. It had a wood stove for heat and cooking, and I had the chore of chopping the wood. Each morning I chopped and chopped and chopped, and then one day I took off a portion of my thumb; my thumb is still kind of flat on the end. But living out in the woods did have some advantages. We were close to the Umqua River, which Chuck and I could explore, and there were hills to climb, and we had a little cocker spaniel puppy to play with. There was always something good about every place we lived, and I enjoyed life with the family.

We were only in Oregon a couple of months, but this was long enough for the rains to start. It rained as it can only rain in Oregon. The river came up and looked as if it were going to come in the back door. There was too much rain, no sun, and a very poor job for Dad, so we decided that Roseburg, Oregon, was no place for the Irwin family. Dad went off to Salt Lake City, looking for another home, and, in a few days he was back—he had found just what we wanted. Salt Lake City was the place for us.

In late fall of 1944, we arrived in Salt Lake City, and I enrolled in the Horace Mann School. Dad had a job as a plumber with the University of Utah, and he moved us into our first home there, a single room in a small hotel on South Main Street. We were directly across the street from Z.C.M.I., a department store owned by the Mormon Church. Since we didn't have any kitchen in our little room, we ate all of our meals in the "grabeteria" in the hotel.

Mom and I had a little excitement on one of our downtown Salt Lake strolls. We were walking west toward the train station one morning when I saw smoke pouring out of a window on the upper floor of a building. We rushed upstairs and found a woman and a small baby in a smoke-filled room. I remember that Mom grabbed the baby, and I started pouring water on the fire with a dishpan. We got that situation under control—my first attempt at being a hero—and there was a writeup in the *Deseret News*.

We bought a house out on the east side of town on what they call the Bench, at 1903 Blaine, and I transferred to Irving Junior High School. The Bench was right at the base of the Wasatch Mountains. During the years in Salt Lake, one of the great joys of life was climbing those hills and mountains in summer and skiing the valleys and the canyons in winter. From our home in the east foothills, the whole family could climb Old Hunchback, which was directly behind us. This was an important part of our lives, and it gave me a deep and an everlasting love for the mountains of the west.

Beautiful canyons surround Salt Lake Valley—Immigration, Parley's, City Creek, Big Cottonwood, and Little

Cottonwood. There were fascinating hikes through these canyons, and excellent skiing in the winter. At the head of Big Cottonwood, there was a charming little community called Brighton; and from all of the major canyons there were trails leading up to the mountain lakes and to the heights of the peaks. Our outdoor lives brought us tremendous exhilaration, and occasionally a little adventure.

One April, Dad and my brother and I were hiking up to Lake Blanche from Big Cottonwood Canyon. We were having tough going through some snow when we heard wild cries for help. A woman had injured a leg falling off a rock ledge, and we very luckily came upon her. I put my first-aid training to work—wrapped up her ankle and improvised a drag device like a sleigh. After a really exhausting struggle we managed to get her down the mountain. Turned out she was a high-school teacher named Irene Mulholland. After she had received proper treatment, she told her story and I got a writeup on the front page of the second section of the *Salt Lake City Telegram:* "Salt Lake Boy Saves Teacher after Mountain Fall." Here I was, twice a hero in Salt Lake City. They really kidded me in school.

The second summer I tranferred from Irving to East High School. About this time, my mother met the Woods family. Bill Woods operated the shoe department at Makoff's, an exclusive women's store in Salt Lake City. Mother convinced them that her son James would be the best yard boy that they could possibly find. I worked in the yard, cut the grass, did odd jobs, and occasionally helped Bill Woods with the rock walls he was building around his property. These were walls of great beauty. Bill is a superb artisan, a man skilled in many crafts, and every morning he would get up early and go out and get his exercise by building rock walls.

One morning, Bill asked me if I'd like to come to work

downtown at Makoff's as a shipping clerk. I jumped at the opportunity. Soon I knew Mr. Makoff and his two sons, Richard and Sam, and I met Ed Spitzer, who really ran the store. The group at the store became my second family, and, in a way they brought me out into the world. It is amazing how Providence works to accomplish ends that seem beyond possibility.

In Salt Lake City I was very fortunate in my teachers. I remember many teachers there, but my great favorite was D. E. Powelson, who taught physics. He really developed my interest in the subject, and this made me feel differently about all science from that time forward. Mr. Powelson made the students work. The lab work in physics was so interesting that it helped me later at the Academy and in graduate school.

Dad wanted me to sign up for ROTC in high school. I did, but it wasn't a very auspicious military beginning. I ended up as a sergeant in my senior year—all my friends were officers. There was one innovation that I thought of. Instead of shining my shoes, I decided to shellac them. It looked good for a few times, but when it began to crack it was a real mess.

I was pretty much of a drudge in high school. I was studying hard, working at Makoff's after school, and doing a little yard work on the side. I didn't have time for athletics or dating. I dreamed about girls, but only wondered if I would ever have a girl friend of my own. Dad kept me abreast of the recent developments in infectious diseases and alert to the possibility of losing my money.

My folks decided they needed a roomer to help out with the finances. They scurried about and found a candidate; amazing as it seems, it turned out to be a girl. Evidently this young lady was having problems with her family in California, and it was arranged that she would come and live with us for a while. Chuck and I moved out of our room, and Dad fixed up a place in the basement for us. It was fairly nice—and it was also right under our old room, now hers.

She turned out to be a shapely brunette just about my age. Here I was, who had never had a date, and with all these fears and desires and wonderings about girls, and one was right above me, the other side of half an inch of flooring.

Well, there was this large pipe that came out of the furnace and up into a ventilator in her room. I told Dad that we were freezing to death down in the basement, and that we had to have a hole cut in that vent. He agreed, and I cut the hole. I had somehow managed to get my hands on a periscope that was just the right length to stick up the vent and look into her room.

I will never forget the night that we got everything set, and I got the periscope in place. But we started to laugh, the periscope rattled around in the ventilator, and she closed the ventilator in our faces. After all that engineering work and planning, I blew it.

I was very romantically attached to that girl, and I would have loved to date her. It broke my heart, but it was never meant to be. My best friends from high school came over and took her out. It seemed everybody dated her except me.

During those years I was a pretty kooky-looking guy. I was freckled and my hair stood up in the back. People called me Freckles. You remember that guy in Our Gang, the skinny one whose hair stuck straight up in back, Alfalfa? I looked so much like him that people used to stop me on the street and ask me if I were Alfalfa. Frankly, I was kind of embarrassed by my looks, and I never thought any girl would be interested in going out with me. That plus my father standing on my back really killed any initiative I might have had getting dates. Looking back on it, I wonder that I grew up at all.

But I did. My first date was a real dramatic event. Even my dad was proud. I decided that if I was going to have a date, I would go after No. 1. The occasion was the annual Christmas party at Makoff's. The year before I had not taken a date to the party, and I was not happy with that. This time I asked Jean Maw, the daughter of the governor of Utah, if she would go with me. She said yes, which surprised me. I think she was close to going steady with somebody there in high school, but she agreed to go. She was a very good-looking girl.

Of course there was no question as to what I would wear. I only had one suit, an old green one. I asked Mother if she would press it for me. Doggone if she didn't burn the collar, a big brown spot. If you can imagine it, she took the coat collar off and reversed it to hide the spot. Here it is my first date, and I am going out with the governor's daughter, and I am wearing this ancient suit with an upside-down collar.

It was not a very relaxed evening. We danced, and I had never had any dancing lessons, but I do not think I stepped on her feet too much. I don't know how I got her there and back—we must have taken a cab. I really did not even know what to talk to her about, let alone whether or not I should kiss her. I think I did kiss her good night when I finally took her back to the Governor's Mansion. If so, she was the first girl I ever kissed.

I never will forget how Jean looked. She had long blond hair, a turned-up nose, and the sort of Teutonic good looks that I have always been attracted to. A short time ago, I went back to the twenty-fifth reunion of my high-school class, and

Jean was there. I enjoyed seeing her again. She looked amazingly the same.

All the while I was finishing up high school I wondered what I would do next. Where would I go to college? Wherever I went I would need financial help, because my family couldn't contribute much to my education. I was determined to be successful in school and on the job. I had gotten straight A's the last two years, and I had learned a lot at Makoff's. With Dad's encouragement, I was still dreaming of going to West Point. Finally, with a little help from my friends, the problem of which direction to take was resolved.

My dad wrote to Senator Elbert D. Thomas and told him of my interest in West Point. It was a careful handwritten letter describing my school record, my work down at Makoff's, and what Dad considered to be my qualifications. Dad is a good writer—he has the ability to put his thoughts together in a beautiful way. My big boss, Mr. Makoff, wrote the senator, too, and told him what he had observed about me and what Bill Woods had told him. Beyond this, Mr. Makoff sent me to a Mr. Morris Rosenblat, who owned a steel plant in Salt Lake City. I was a little apprehensive as I rode across town on the streetcar to see Mr. Rosenblat. But we had a friendly conversation, and he was impressed enough to write a letter of recommendation to the senator.

The senator got at least three letters about young Jim Irwin, but he was not about to propose me until he knew that I was qualified, so he arranged for me to take a substantiating examination. Out of a possible 4.0, you had to get at least 2.5 in all subjects. I squeaked through with a 2.54 on English, but I did a little better in the other subjects. West Point was my selection, my desire, but I was quickly aced out, and the only possibility for me was an appointment to the

143

Naval Academy. Not too many young people in Utah are interested in the Navy. I went for it, was recommended by the senator, and was lucky enough to get the appointment.

I didn't know anything about the Navy, but Dad had pointed Annapolis out to me from the deck of the ferryboat and said, "Son, maybe you will go there some day." He was right. Here was my college education in hand, and it was a great honor. The appointment came through about two months before I was to graduate from East High School. I was scheduled to report to Annapolis in early June. When Dad got the telegram, his eyes lit up and the load dropped from his shoulders. Mom was so pleased she immediately began planning to take me back east herself. She couldn't let go of her son.

Bill Woods was a great influence on my life during the years in Salt Lake City. In the evenings, Bill would give me a ride home and talk to me about life. He had a broad, expansive understanding of the world we live in. He was a worldly man, and yet a very devout Christian, and he encouraged my interest in the church.

We actually went to many churches in Salt Lake City. It's strange that we always seemed to be shopping churches. Since we were in a Mormon community, I attended the Mormon church a number of times. But there wasn't enough talk about Jesus Christ and too much about the Book of Mormon. So I went off to Annapolis without having finally settled on any one church or denomination.

After high-school graduation I had only one week to get ready for the Navy. There was terrific excitement around home, but since I didn't have to worry about getting a wardrobe ready (that charred green suit was my wardrobe), packing was no problem. Mom and I took a train from Salt Lake City and headed east through the Royal Gorge. For some reason we were routed through Colorado Springs. We got off the train and spent the night there. It was the first time I had ever seen this beautiful little town. I vowed that if I ever had the opportunity I would move to Colorado Springs because it was the most beautiful little town I had ever seen.

Next morning we caught a train for the East Coast. Let me say Mom is great to travel with, but I was beginning to feel a little bit like a mama's boy. It's embarrassing for your mother to have to take you to Annapolis—to deliver you at the gate. That's exactly what she did. She turned me over to the Marine guard at the gate and said, in effect, "This is my boy Jim. He's a good boy and I want you all to take care of him." I was eager to get away from her, and I walked right through the gate and into the Naval Academy.

Immediately I was sworn in and taken to a barber shop, where they sheared off most of my hair. This was the first professional haircut I ever had in my life—Dad had always cut my hair. Then I was given a little uniform called "white works," which I had to live in for the next four years. Within a day or two I longed to be back home with the family. But, man, it was a whole year before I saw them again.

6

MILITARY LIFE

I AM SURE I was wearing that green suit when I arrived at Annapolis; now it had seen me through the second great event of my life. I said they immediately cut all my hair off, and it wasn't particularly long. I guess this is one thing I have always objected to about the military. I must have always been a frustrated long-hair; at least I've always wanted it longer than they would allow.

They assigned us roommates. I got Bob Bregman from New York City. I didn't realize how wealthy he was at that time. He has a seat on the New York Stock Exchange, and his family has been involved in Wall Street for many years. He seemed much older than I, much more mature. I had led a very sheltered life.

We began a new life together, the midshipman routine. This meant up at 6:15 A.M. every morning, out in the hall at attention. But there were no upperclassmen that first summer, so we had the run of things and it wasn't too bad. They checked us out in sailboats, and we had to pass a swimming test. I didn't have any of the recognizable strokes. So I was assigned to the sub squad and had to go swimming several times a week until I could pass the test.

I was put down good a couple of times. One night after

a big dinner, I substituted for a man in my weight class in a battalion boxing match. They needed a man to complete the card, and I said I'd do it. Well, this redheaded fellow started getting to me; he knocked me down a few times and really embarrassed me in the worst possible way. My first attempt to really shine, and I was laid out and then some. It turned out that I hadn't finished learning my lesson, because soon after that, I challenged a tall, lanky fellow from Virginia named Willy Frasca to a race around the track. He actually had a sprained ankle at the time, and I was really kidding when I bet him $20 that I could take him on the quarter mile. Before I knew it, I was down at the track, and doggone if he didn't beat me. At about that point, I decided that I was going to build myself up, and I have worked out regularly ever since.

All this time I was really missing my family, and I wrote them a letter every day telling them what I was doing. This was almost like talking to them, and it helped a lot. When the upperclassmen came back at the end of the summer, we had a rough time. I was assigned to a different battalion, and I got a new roommate, George Gleason from Teaneck, New Jersey. He talked a great deal, and I was very reserved. I almost had the feeling he was a phony. George was on the swimming team, an excellent swimmer, and I was a lousy swimmer. As it turned out, George and I roomed together all four years and became very good friends.

Being plebes, we were the lowest of the low, and we were constantly being picked at by the upperclassmen. We had to look sharp, answer "Sir," and walk on the outside of the staircases. Of course, we had to eat square meals—that is move our utensils with those 90-degree corners. If I displeased anybody, he would say, "Come around to my room,

Mr. Irwin," and then I might have the opportunity of stand-ing at attention or doing pushups. I was never much amused by these visits.

Fortunately, my mother's sister, Helen Hess, lived in Baltimore. She and her husband took me into their family. They had two children, Herb and Gail, and when Annapolis played football games in Baltimore I would spend the night with the Hesses. Aunt Helen was always trying to fix me up with dates, so I started doing a little dating at the Academy.

I never was one to study very much, just enough to stay pretty high on grades but not as high as I should have been. If you get a 3.4 average or better at the Academy, you are considered to be a Star Man, which is a real distinction. I was always 3.3, 3.38 or 3.39—right on the border. I loved to sleep. I would go to my room after classes, and if I didn't have to go out for athletics that afternoon I would sack out. I'd go back to the room after dinner and study for a while; then I'd go to bed. My classmates couldn't understand how I was making it. Still I managed to get better grades than anyone in my company.

The studies were easy at Annapolis. Most of the courses were memory courses, and no depth of understanding was required. I can remember almost anything for a day or two, so I used a cram-course type of approach. Then, after a few days, I'd wipe it out and go on to something new.

Even after I became a first classman I was not free of the system. I was lying there asleep in the sack one night and my plebes were over in the room, voluntarily arranging my clothes, cleaning up my room for me, or something—doing something they shouldn't have been doing. The Officer of the Day came into the room, got me out of the sack, and made me stand at attention. He accused me of making my plebes do my chores. He wouldn't believe that I had been sound asleep. He wrote me up on report, and for that I got confinement for several weekends. This shot the weekends—you have to report every half hour at the front office.

I was pretty well incensed by the Naval Academy, and this incident was the culminating blow. By this time I had given up on the Navy anyway. I didn't want any more to do with it. I had had enough exposure to the life and to the rules. After two experiences of sea duty, I decided this was not a comfortable way for me to live. It seemed ridiculous to spend that much time at sea when there were so many other interesting things to do.

While I was at the Academy, Dad got dissatisfied again. He was still working as a plumber for the University of Utah —which was okay, but not nearly as profitable as if he'd been on his own. He said that it was cold in Salt Lake City, just as bad as Pittsburgh. So the family moved to San Jose, California, and Dad got a job as a plumber with San Jose State College. He surely has never been an entrepreneur or promoter—he has been quiet and industrious—but curiously enough he enjoys a little gambling. Anyway, Mother and Dad have been in San Jose for twenty-five years now. You might think he'd finally found the perfect spot, but he's still talking about moving south.

It was becoming more and more important to me to leave the Navy—I couldn't resolve the Navy life in my mind —when, providentially, a great opportunity came up. The Air Force was created in 1949, and they were looking for young officers. They were allowed to take 20 percent of our class at Annapolis, with the volunteers to be selected by a lottery. I drew a number less than 100, and I elected the Air Force. If I had been stuck in the Navy, I would probably have left the service in the minimum time.

We graduated from the Naval Academy in June, 1951,

about three weeks after I had broken up with a girl who had assumed that I would be taking her to all the dances and festivities. So, I didn't have a June Week date. I took my mother to a couple of the affairs and she enjoyed them—so did I.

That summer when I left the Naval Academy, I had about two months off. It was a great vacation. I was in the Air Force as a second lieutenant, and I was getting more money than I had ever had in my life. In fact I had so much money that I went off and bought a little red coupe, a '48 Ford. It was only three years old, but it seemed older than that. It cost me $800 of the $1,000 I had saved on my salary of $78 a month at the Academy. Most of the guys when they graduated bought new cars, but I didn't want to go into debt.

I actually bought that car in California while I was visiting the family, and it ran great on the coast, but it overheated as soon as I headed east to report to Hondo, Texas, my first assignment. I remember coming into Hondo early in the morning just as the sun was coming up, and the air was sweet with that morning smell of the sage and the piñon. It was beautiful country with pretty rounded hills out to the west and north. While I was admiring the countryside, I discovered that I had driven onto the base without realizing it—there wasn't even a main gate. I could see some low rambling buildings off in the distance. I finally found some guys around, and they said this was it. It turned out that they were reactivating Hondo Air Base for the first time since World War II and this was the first breath of life the place had seen—human life, that is. The buildings I spotted had been used as chicken coops. They still had chicken wire up, and there were feathers and droppings all over the place. It was the most primitive living I had ever seen.

Within the next few days, most of the other guys filtered in, and we were divided into groups of four or five and assigned to instructors. We were living barracks style, sleeping on metal cots, and with one bathroom for the whole group. There were some great guys—we had a combination of Annapolis and West Point graduates. Our instructors were civilians under contract to teach us how to fly. My instructor was Ed Siers, a colorful little guy from south Texas who had been a crop duster and had done all sorts of flying. He saved my career.

Our group was full of generals' sons and top graduates from Annapolis and the Point, and competition was really keen. We knew when we went into flying training that some of the guys were going to wash out. It turned out to be 50 percent of my little group, an unusually high percentage.

I wasn't particularly interested in flying and I wasn't pressing at all; I was hanging loose. Looking back on it, it seems rather ludicrous that I should have made it. As I have said, I actually told my instructor I didn't feel comfortable flying, and he sent me to the Base Commander, who sent me to the Commandant of Students. But I couldn't get out of pilot training because I wouldn't say that I was afraid to fly. I wasn't afraid; I was bored. So I made the best of it for a while.

Christmas leave came up, and I had a great time skiing with my brother, Chuck, at Badger Pass near Yosemite. We were enjoying the beauty of nature in those magnificent mountains, and everything was lovely until I started back. About Southern California, I began to feel mighty bad; we pulled into Tucson on New Year's eve, and I was lying down in the car feeling horrible. The other guys went into a local club and partied all night and then we drove on to Hondo. I was just barely hanging in there, and the next morning, I just made it to a telephone to call the hospital to come get me.

I had viral pneumonia; my fever was so high that I was out of it for a couple of days. My instructor, Ed Siers, came by to visit me. There I was flat on my back and losing precious time every day, getting farther and farther behind my class. Ed became a great friend. He encouraged me to get out of the hospital and catch up. He even volunteered his time to fly with me on weekends. Somehow this crisis got me motivated, challenged me to catch up with the class and graduate.

Everything worked out, and I took my T-6, that funny little yellow airplane, and soloed. I wasn't uptight. It was a great joy—I mean complete relief—to be up there by myself. I realized what a loner I am when I got the great charge of exhilaration from being absolutely alone up there in the sky. When we came back from the first solo flight, they hosed us down with the fire hose. That was a happy day. The next high point was the first night cross-country flight on instruments. I remember once being a little disoriented and wondering if I was looking at city lights or stars—they were city lights, and I was relieved when I got on the ground at Hondo.

No sooner had I stopped fighting it than the time at Hondo was up. I was assigned to Reese Air Force Base near Lubbock, Texas, and I thought I knew exactly what I wanted to do. I would make the most of it by going multi-engine. I would fly the B-25's, get all that good training in the big, slow stuff, and then go into airline work some day. As I drove over to Lubbock in my overheated red coupe, I was building my future in my mind.

When I got to Reese Air Force Base, there was a whole different deal. They were looking for little guys to train as interceptor pilots, and they took all the little guys (including

me; I'm five foot eight) out of our class and put them in
fighters. We were going to be put through a special instrument course that would make us absolutely great all-weather
instrument pilots.

Translated, this meant that we were going right back to
the little yellow T-6's that I had just left with such relief
back in Hondo. We never flew anything at Reese except the
T-6's. I was a little embarrassed to accept my wings after 230
hours, every one of them in a T-6.

The plane was the same, but Reese itself was very different from Hondo. There were many permanent buildings
and the barracks were two-story chicken coops, but there
were rugs on the floor and we had individual bathrooms. My
new roommate was Wallace Christner from the Naval Academy. He had red hair, buck teeth, and a round face, and I
think he was a farm boy from Ohio. At this point I felt I
could get along with anybody.

It turned out that Wallace was going to play a fateful
role in my life. He introduced me to a striking brownette
girl with a vivacious manner named Mary Etta Wehling.
Flying at Reese was dull, no challenge, but I didn't have
time to think about flying. I fell in love with Mary Etta. She
was all challenge. Flying was incidental, and seeing Mary
Etta was everything. Once I met Mary, there wasn't any
other girl that interested me. I saw her almost every night of
the last two months of Advanced Training.

Mary Etta's father was in charge of maintenance at
Reese Air Force Base, a senior captain who worked his way
up through the ranks. Since I was twenty-two and she was
eighteen and a senior in high school, he was a little uneasy
about me and a little standoffish. He was a devout Catholic, and she had been raised in the Catholic church. This

religious problem was something we couldn't escape because of the intense anti-Catholic sentiment in my family. My whole background had conditioned me to believe that there could be no satisfactory marriage with a Catholic girl.

I kept telling Mary that if we could work out this religious problem we could go ahead and get married. I was trying to convert her, and she didn't want to hurt her father, a strong influence in her life. I felt that until we solved the religion problem, which actually meant until she gave up her religion, we couldn't consider marriage. The time had come for me to pick a base. I wanted to be near my parents, on the West Coast, and near mountains I could ski and climb, so I picked Yuma, Arizona.

Nothing was resolved with Mary, except that she knew I was going to leave and there was no way of knowing how long it would be before we'd get together again. There was a big graduation ceremony with a parade and a fly-by and everything, and Mary pinned my wings on. She was the prettiest girl there, as far as I was concerned.

About this time I turned in my '48 Ford and bought a '51 Kaiser, gray on the outside and with pink upholstery. Looks like I'm following in my dad's footsteps. I said goodbye to Mary, and then I went home for a couple of weeks of vacation before reporting to Yuma. Of course, I had conversations with my family about marrying a Catholic girl. I had already written to them about it, and it had doubtless been discussed before I got home. They thought it would cause a problem for me.

After the brief vacation, I drove to Yuma. It was a stark desert scene with tumbleweed blowing across the road and sand in the air, but fortunately I had a feeling for the desert and would grow to love it. It wasn't an interceptor squadron at all; it was a gunnery base. But I saw some P-51s there on the strip, and I mean that changed my whole feeling. I had not flown anything but a T-6, and those 51s were the hottest planes I had ever seen in my life.

A Major Dumont was the squadron commander, and he met me with a very large German shepherd at his side. The major was diffident yet friendly. He took me over and introduced me to the guys in the squadron. They were a grizzled bunch back from Korea. Some of them had shot down MIGs and had flown 51s in combat; most of them had flown 84s and 86s. They met me pretty coldly.

My new home was a rambling wooden building up on the hill that must have been the old officers' quarters back during World War II. It had three small rooms and a piece of a long porch. I sat there at night, a hundred feet up on the ridge, and looked out over the base. To the north you could see some wild-looking mountains: Picacho Peak sticks up like a giant finger out of the desert, and to the east of that beautiful spire is Castle Dome. Both are part of the ranges of mountains that run north and south along the Colorado River.

They checked us out in the old T-6 first, and we made the type of landing where you touch down with the main wheels first, in a level attitude so that you can see the runway, and then gently lower the tail. This was the technique that we would use for the 51, which packed that mighty engine into a long, pointed nose that cut off your view when you were landing or taxiing on the runway. You had to feed in almost full rudder to keep the 51 straight going down the runway, or the torque might cause it to ground loop, or spin on the ground. After we satisfied them that we could make the right sort of landing and were basically familiar with the plane, they checked us out in the 51. It was unbelievable—not related to any flying I had ever done before. The 51 was one of the greatest airplanes I have ever flown, and I can say this retrospectively, having since flown almost every airplane in the Air Force.

From that point on, I found myself living to fly. It was a consuming passion, and it took me over. Our job was to tow targets for all the interceptor squadrons; each squadron would have gunnery practice about once or twice a year. I quickly gained the reputation of flying more tow-target missions than anyone else. In fact, some of the old hands who were veterans of Korea thought it was beneath their station to tow targets, so they would give me their missions. Almost every month I had over a hundred hours of flying time.

Our targets were 6 feet by 30 feet and were towed behind the airplane on 1,000 feet of armored cable. It required a little technique in taking off, the point being that you tried to preserve the target. The cable would be snaked down the runway with the target lying behind it. You would gun the engine, release the brakes, and, just as you took up all the slack and the cable went taut, you would try to take off and pull up as straight as you could without stalling out. If the engine conked out, you would have had it.

Our base commander was Bob Worley, a full colonel and a great man, who was responsible for starting up this new gunnery operation. The Air Defense Command was transitioning into jets like the F-89, the F-94, the F-86, and on into the Century series: the F-101, the F-102, and the F-106. I would be involved for a long time in the struggle of the Air Defense Command to get new interceptors.

As soon as I got the required hundred hours in the 51, I made a cross-country flight back to Reese Air Force Base to see Mary. I had to take Captain McCurdy, a flight leader, with me and fly his wing on the checkout flight. After refueling at El Paso, we took off and flew right through an Army gunnery range. Of all things—he was supposed to know where this range was. I remember seeing the bursts of antiaircraft fire off the wing and taking evasive action to get out of there. When Captain McCurdy dropped me off at Reese I was so excited I did two 360s over the field (circled twice in fighter-plane style). Mary and her father were there to meet me. I had a great weekend. Then Yuma sent a flight leader to pick me up, and we flew back to my new base.

Colonel Worley treated me well, and he became my hero. He asked me to come up and talk to him any time I felt like it and tell him how things were going. He seemed to have a special interest in the Annapolis boys, and we developed real rapport. He went on to get three stars, and then he was killed over Vietnam flying an F-4 on a raid. Got a direct hit and just blew up.

I really missed Mary, and we started writing. I must have written her every day. About this time she graduated from high school and started going to Texas Tech. I still pursued her, telling her there was not any hope as long as she was a Catholic. Finally, she started to soften up and got to the point where she said she was going to give up her faith. She said she was going to come and join my church, that she missed me so much there wasn't anything that she wouldn't do for me. We planned a December wedding.

Looking back on it, I guess I couldn't stand being told what I had to believe—and yet it seems I had to insist on what Mary should believe. Anyway, I was headstrong and it was hard to change my mind. On reflection I have to admit that stubbornness is something of a family trait—Dad is stubborn and Chuck is stubborn and Mother always had her own ideas too.

The weekend came, and I drove up from Yuma to Reese. We got married at the base chapel. Mary wore a white dress and a veil, her father gave her away, and her brother was my best man. My mother was the only member of my family who came for the wedding—I guess it was a little funny that Dad and Chuck weren't there. Of course, we didn't have much money.

After the wedding, Mary and I jumped in the car and headed for Yuma. Actually we headed for Ruidoso, New Mexico, for the honeymoon night—this is west of Roswell up in the Sacramento Mountains. We were riding along in my old Kaiser with the pink upholstery. When we got there we checked into a nice little motel. I was twenty-two years old, and I had been waiting for a long time.

Early the next morning, we struck out for Yuma. I didn't want to take any time off; I wanted to get back and fly. We had a little apartment in the midst of dense tropical foliage, out in the west end of town. There were four units around a swimming pool. Most of the tenants were young married couples, and we had a great time from the first, great parties. I was making about $500 or $600 a month as a first lieutenant, and I think the apartment cost about $100 a month.

Like most of the other officers' wives, Mary lounged by the pool all day, played cards, and got a terrific suntan. And I was still flying real hard; marriage didn't affect my flying. If anything, I was more relaxed. The squadron really delighted in our marriage, and they accepted Mary easily because she was a fun gal to be with; she fitted right in.

I never will forget a masquerade party, probably on Halloween, when I went as a mummy. I got this muslin target cloth and cut it into strips about two inches wide and had Mary and the neighbors wrap me from head to foot. I was totally wrapped, with little slits for eyes, and I won first prize. I felt so good, and was so elated, that I started drinking and really hung one on. I can recall going up to receive the prize and falling flat on my back, right on the stage. I also remember trying to get Mary and me to the car afterwards. I was trying to carry Mary, and I fell down right outside the officers' club where there wasn't any grass, just a lot of sand burrs. I began picking them out of my skin, as I staggered on —finally we made it to the car.

Because of our immaturity we argued about the most trivial matters. Our first real fight started over something silly like my picking the most scenic way for a trip. She said, "You always pick the way. Why don't you let me?" So we fought and she went back home to Lubbock for a couple of months. Finally, she wrote and said that she would like to come back to me, that she thought our differences could be resolved. I said fine, come on home. I was unhappy without her, I was married to her and loved her.

At about this point I began to transition into jets—T-33s —and I was fascinated with the flying. They were reorganizing the base, it was growing and expanding, and the interceptors that we were towing targets for were also jets, firing guns and rockets. We had gone from the 51s to the T-33s, and finally we would move on to the B-45s for tow planes, which were the first jet bombers that the Air Force ever had. Now our targets were 9 feet by 45 feet and the cables were longer and our tow planes had the capability of staying up a lot longer. We didn't have to tow those targets off the ground; we released them from the bomb bay.

I was the first fighter pilot on the base to make the transition to the big bombers. The biggest thing I had ever flown was a T-33. To get into this monster, the B-45, you climb through a door, walk down a bit, then go up a flight of stairs to the seat. You look out over these huge wings and the banks of instruments and controls and wonder if you will ever be able to learn all that stuff and fly the thing. After you get off and go around the pattern a few times, you notice you can operate four throttles as well as you can operate one, and it lands and takes off like a T-33. It has a wheel instead of a stick, and instead of flying it like a fighter with the stick in your right hand and the throttle in your left, the B-45 is just the reverse; you have your left hand on the wheel and your right hand down on all those throttles.

I switched squadrons from the tow group to a training squadron which was Major Jabara's outfit. Jabara was our first triple ace, and he was a pretty difficult guy. He gave everybody such a hard time—and he gave me a hard time on a couple of occasions, too.

During this whole time Mary and I were not going to any church. I can't imagine it, but I don't think we had any church relationship of any kind. There I was telling Mary she could not be a Catholic, but I was not giving any spiritual leadership. I believe there were times when she was thinking of going to visit a priest or just going to a Catholic church. We obviously had not solved our problem, and it was coming out into the open again. Looking back now, I don't believe that we were communicating with each other in any deep way.

Then, in the summer of 1954, we went down to Moody Air Force Base in Valdosta, Georgia. I was there for F-89 school, and Mary and I and our cat rented a house. (Mary had brought this cat back from Lubbock the time we had a fight and she had gone home.) It turned out that Mary had lived in Valdosta when her father was stationed at Moody years before, and she knew the Catholic priest. One day she told me that she had gone down to see him. It really torqued me off. I couldn't accept that.

I told her that we couldn't continue if she was going to do that, so I sent her home. Just like that. It was mighty cruel on my part, looking back at it. But she didn't argue with me, didn't talk back or anything. She said, "Okay, if that's the way you want it, I'll go." And she left. She left so quick that the cat was left behind, and I had to find a box and send the cat on a couple of days later.

After Mary went home, her father helped her get a divorce in Texas. A little bit later I found out that I was divorced, so I gave up the house and moved into the Bachelor Officers' Quarters (BOQ). Looking back, it really makes me seem like a cruel, cruel husband, to drive a wife out with no more effort toward reconciliation than that. In trying to understand how threatened I felt by her efforts to contact a priest, I guess that I resented the authority of the church. Maybe I feared the church would have more authority over her than I had. I never had any communication with Mary after I sent her back. But about seventeen years later I took my present wife Mary and the kids over to see Mary Etta's family in San Antonio. They were very warm to us and seemed genuinely glad to see me and meet my family.

I was in the F-89, the old Scorpion, at Moody. It was an ugly-looking plane with two intakes underneath and a high tail. It looked like a pregnant spider with a snout. It wouldn't go very fast, but it could carry about 104 rockets, a tremendous armament load. On the F-89, you were assigned a radar observer to operate the radar, pick up the enemy planes coming in, and prepare the rockets. The pilot fired the rockets. I had this little guy named Doug Anderson from up around Brainard, Minnesota, as my radar observer, and we became good friends.

I loved flying, but I could see that if I really wanted to progress in this aviation thing I would have to get some more education. I'll tell you what triggered me. One day we were sitting out there on the base, and a couple of guys came down and made a slow pass over the field. They were out of Edwards Air Force Base, and they were flying 104s or 102s— you know, these real sleek, advanced jets. I knew that if I was going to be a test pilot, I had to go to graduate school. And I decided I'd take the initiative and tell the Air Force where I wanted to go to school rather than waiting for them to tell me.

I picked the University of Michigan, because I had some good contacts with guys stationed over at Hollaman Air Force Base, and one of them, Buck Buchanan, had gone to the University of Michigan. Buck had come over to Yuma to fly our F-89s, and I was his instructor and checked him out. He had recommended a course in Guided Missiles. So I made a formal application to my command, and I got a positive response.

Early one morning in the summer of 1955, I said my good-byes and headed out of Yuma, driving my apple-green Thunderbird convertible—I had gotten rid of my Kaiser and traded up. I was particularly glad to get away from Yuma because I had had one last bad experience with Major Jabara, the triple ace.

On a rocket-firing mission they always assign a chase plane to follow the plane that is doing the firing, just to observe and to clear the plane to fire. At a given point, you have to make the decision as to whether you are going to fire or not—if you elect to fire, you press the trigger and that commits you. The chase plane has the responsibility of break-ing it off if he thinks the rocket may hit the tow aircraft in-stead of the target.

I was the chase plane on Major Jabara's tail, and I cut him off. I was certain the big ace was locked onto the tow plane rather than the target. He gave me a hard time on the radio, and then when we landed he chewed me up one side and down the other. And he did it in front of all the guys in the squadron. It was pretty humiliating.

Leaving Yuma behind me, I was rolling it around in my mind. I thought it had been a sad time and it had been a happy time. Yuma had been a real consciousness-raising place for me: I had developed a love of flying, and a love for Mary, and a love of the desert. I had had some wild ex-periences and some sweet experiences. I had cut some capers, and I thought about them as I cruised along at 75 miles per hour (1,500 rpm). Many times when I had made that run in the 51 from Yuma to San Bernardino to Palm Springs, I'd drop down and fly the railroad track, five to ten feet off the ground, in hopes of finding a train to scare. The engineer would look out and see an airplane coming down the track with my head on the level with his head. Or we'd get two or three planes together and chase each other down the Colorado River, five to ten feet off the deck, scare fishermen out of their boats, and churn the water with our propwash. That was behind me.

I had never been to Ann Arbor before, and I found the country to be beautiful, with lots of lakes and a warm after-noon sun on the hills. When I got to the university I looked around the campus; I have never seen so many ugly girls in my life. I asked about this, and somebody said that all the good-looking girls were up at Michigan State. But the land was lovely in late summer—a lot of green, with leaves whispering in the bright carbonated air—very different from Yuma. Immediately I started reading the ads for a room, and it wasn't long before I found the Edgar Johnstones' home, out on the east side of town in a fashionable suburb. I had a nice bedroom and bath over the garage, looking out on a beautiful back yard.

I had everything well under control, except one vital consideration: parking in Ann Arbor was impossible. There was no way I could run around town in my T-Bird. So I bought an English racing bike. I used it year round, even when the snow was in four-foot drifts. I got a lot of fresh air and great exercise, and it reminded me of the bike that I had bought from the mayor's son in New Port Richey.

No sooner was I settled in than I began to miss flying, so I started scrounging around. I checked in with Selfridge Field. They had a C-45 for me, a miserable little two-engine airplane with a tail wheel. It was very difficult to take off and land, and as I checked it out I noticed that they had a couple of squadrons of interceptors there; they looked like F-86s. Actually, it was a reserve squadron, and they were really flying F-80s and T-33s. But this really made me ache, and it didn't take me long to get over there to talk to the operations officer.

He was Captain Kuntz, and they called him "Black Scot." I told him what I'd been doing in Yuma and what I had been flying and said, "I'd love to fly with you. Do you think you could possibly use me?" He said, "Jim, maybe we could use you." Man, this was the answer to a prayer. "We don't often do this, take people like you into the squadron," Scotty said, "but we need a little help. We'll check you out in the T-33, and maybe you can fly the F-80." I had never flown the F-80, so this was a great opportunity. I'd get up every Saturday morning about 4:30 A.M., jump in my T-Bird, and drive to Selfridge, where I would work with students all day. Soon we transitioned into F-84s, another plane I had never flown, and it was absolutely great.

During my time at Michigan, I was really studying sixteen hours a day during the week. This academic program was supposed to prepare me to do management work and research in guided missiles in the Air Force. I was taking basic courses in aerodynamics and structures and advanced courses in electrical engineering, aeronautical engineering, and calculus. This course work was designed to lead to two degrees, a master's in aeronautical engineering and a master's in instrumentation. We were studying surface-to-surface, surface-to-air, and air-to-air missiles.

In a course in advanced calculus I got an E on my first test, the lowest grade in the class. This really shook me, and I buckled down. I got an A-plus, 100 percent, on the next test, and I finally ended up with an A in the course. I studied hard enough to hold a B average so they wouldn't drop me out of graduate school.

There were some great friends of mine at Michigan at this time. Jim McDivitt, who went on to the astronaut program and became the Apollo Program Director, was there, and so was Ed White, the astronaut who later was killed on the launch pad during a cabin test for Apollo 1. He and I used to have breakfast together and talk about what we were going to do with all this education. I was a year ahead of these guys, so this was a real pressing question for me. I had been inspired to come to Michigan in the first place by those test pilots flashing over the field at Yuma, and I still was thinking about becoming a test pilot. I had the theory and the flying experience, and I thought it would be ideal to combine the two and go to test pilot school at Edwards Air Force Base. So I put in my application.

The Air Force said yes, and man, I tell you, that was one of the happiest days in my life. Everything was absolutely perfect—and then it happened. I got a notification from the Air Force that they had canceled my orders because of a change in policy. A ruling had just been made that no Air Force man could go to two schools consecutively. I was thunderstruck.

Where were they going to send me? They were going to send me to Wright-Patterson Air Force Base in Dayton, Ohio.

This was probably the worst place in the world that they could have sent me: bad weather, bad location, bad for my career. I had no wish to go there. I was going to a desk job. I was really sick at heart. Looking back on it now, it was ironic, because the next year they changed this policy and Ed White and Jim McDivitt were allowed to go directly to Edwards and right on into the astronaut program.

Meanwhile, my brother, Chuck, had joined the Air Force and was flying multiengines at Vance Air Force Base at Enid, Oklahoma. I had an afternoon free, so I went out to Willow Run, took one of their F-84s, and flew down to visit with Chuck.

When I was ready to take off and head for home late that evening, the strut was flat; that means the cushion of air had leaked out of the landing gear. And there was a new factor due to a change in weather. It suddenly became an Instrument Flight Route rather than a Visual. I had to run back into Operations and tell them to change my flight plan; it was going to take a little longer. I didn't allow the proper paper reserve that we are required to have on a flight plan, a minimum twenty minutes' reserve fuel. Paper work turns me off; I don't even want to think about it, and I didn't that night.

I jumped into the airplane and flew back to Willow Run, and it turned out that somebody in Operations looked at my flight plan and processed a violation. Where did they send that violation? They sent it to my next base, Wright-Patterson. So, when I reported to my new base, the first thing that hit me was a flying violation. Within a week I was reporting to General Haugen. He said, "Captain Irwin, we are glad to have you in our organization, but it is unfortunate that you are reporting in here under a cloud. You should give more consideration to your responsibilities as a pilot. I think you are rushing just a little too fast."

I checked in with my boss, Horace Maxey, a major, about forty-five years old with a red, fleshy face, balding and about five feet six, yet dignified in manner. He was an engineer, not a pilot. Horace told me that my job was to be project officer on a new missile that was called—well, it was so new it didn't have a name yet. We called it the GAR-9 —that is, Guided Aircraft Rocket 9, really a super Falcon missile. Now, the Falcon missile is an air-to-air guided missile that they were using on the really hot 100 series: the 102s, 106s, and even the 101s. But the Air Force needed a missile with increased range and performance, a nuclear missile. We had the task of developing it.

I heard about an apartment that was available south of Dayton on Alexander Belbrook Road. I drove down the next day and found that the address was a little estate. It was in a valley with a little stream running through it between big wooded hills. You wouldn't believe the view. I was wondering at the many trees and the landscaping, and then I realized it was a golf course. I found the owner, Shelly Lewis, and he took me down and showed me a basement apartment. There was an open-air entrance in the back, with a tree growing in it, and you had to kind of walk around the tree to get in. Once inside, there was a big window, so you could look out and see the tree and the sky.

I walked into this large, comfortable living room with a huge fireplace. There was a big bedroom and bath.

Shelly said, "You can have this for forty-five dollars a month, completely furnished."

I said, "I can't believe it. I'll take it."

So began Jim Irwin's life in the wildwoods. Well, I spent at least two weeks down on my hands and knees scrubbing that place up. It was thick with filth, just cruddy. I was determined to be proud of it.

Wright-Patterson was a complex of engineers, and that means there were a lot of secretaries, a lot of pretty girls. I was back in the bachelor group now, with a couple of footloose classmates, so we proceeded to carry on a very active social life. Chuck was now a multiengine pilot assigned out near Moses Lake, Washington, in the middle of some great ski country—so that meant some wonderful weekends for me. Wright-Patterson was not hard to take.

Looking back, I realize there was only a flurry of social activity before I became preoccupied with making a connection with flight operations. I did and they made me an instructor pilot in the T-33. I was off and flying again; almost every weekend I had a student, there was an airplane available, and away we went. I didn't even see much of my apartment. Of course, there was a lot of traveling involved in visiting the manufacturers of the components for the missile. Hughes Aircraft at Culver City was the prime contractor, so I spent a lot of my time in California. I was mostly there during the week. If I met a good-looking secretary or office worker, I usually tried to arrange a date with her in the evening.

It quickly became pretty clear that the missile was being developed for a new interceptor, the F-108. They merged the planning for the fire-control system of the plane, the missile system, and the interceptor itself into one office which they called the F-108 System Project Office. This meant that we moved into another building and all worked together, with North American, who had the contract for the aircraft, being very much involved. I still managed to fly a lot with my students in the evenings; I was checking them out in the planes, doing proficiency tests, cross-country checkouts— anything they needed. This was very important to me, because the office work was never completely satisfying.

I enjoyed knowing Claire Carlson, the chief man at Hughes Aircraft for GAR-9. But I'll admit that I never did get chummy with Howard Hughes. The work consisted mainly of our being briefed by the contractors and by their quality control people, and taking our assessments of this information to our own guidance and propulsion labs. All this time we were briefing the Air Defense Command in Colorado Springs on the status of everything. They were enormously interested in obtaining another interceptor, because their planes were getting old and out of date.

Some of their staff members would ask embarrassing questions like: "What is the probability of kill if the missile passes a hundred feet from the enemy bomber?" If I didn't know, I'd tell them that I would go back and find out.

The GAR-9, my missile, had a nuclear warhead, so I had to spend a lot of time out in Albuquerque, New Mexico, with the people from the Atomic Energy Commission. Out at the Special Weapons Center we had to deal with such questions as, What kind of yield did we want from the missile? Some of the warhead information was top secret, so I was always working in a tight circle of people who had the same clearance that I did. We had to pick a nuclear device that would do the optimum job: maximum damage, with limited risk to our own planes and to civilians on the ground. If the enemy bombers were foolish enough to come in close, one missile could wipe out a whole formation. Part of our planning was to create a weapon that would change their tactics.

In accord with my usual pattern, I was slipping out of partying into a sort of work-oriented life. But I wasn't a fanatic about it. One weekend I had a T-33 and it turned out that my student was from Seattle. I dropped him off, flew into Moses Lake, and picked up Chuck, and we flew down to San Jose. Chuck had been grounded for six months as a result of breaking his leg in a skiing accident. I had wanted to keep him interested in flying, especially in jets—here was the opportunity to introduce him to the T-33. We cut out that night for a date with a bunch of gals up in Redwood City. I must confess that I drank too much. Unfortunately I had to get up early the next morning to fly Chuck to Moses Lake and then fly to Dayton in the afternoon.

Well, I put Chuck in the front seat of the airplane. I figured that I was a qualified instructor pilot; I could legally do that. We took off from Moffett Field for Larson Air Force Base. I was talking him into the traffic pattern. We did the break and he came around on final, and I told him he was doing all right. As I recall, Chuck actually made the landing, and we touched down on this long 10,000-foot runway. Then we saw this barrier ahead, across the runway. I couldn't believe it at first. I said, "Chuck, man, we've got to stop this airplane fast!" We were both pumping the brakes, and the brakes are really marginal on the T-33. We went right over the barrier. There was a crunch of metal, a whipping of cables. A hangover and here we had zinged an airplane.

"Chuck, we've got to change places. It will be easier to explain if I am in the front seat and you are in the back seat." So we changed places and taxied in. Then I told the tower about the accident and the Flying Safety Officer came out and took statements from both of us. I don't think I ever admitted that Chuck made the landing.

First thing you have to do after an accident is take a physical, so they took me over to the hospital and ran a test. Fortunately, they didn't find alcohol. Meanwhile, my student was over in Seattle waiting to be picked up.

My progress back was ignoble. I was taken back in a little U-3 A, a puddle jumper. When I got back to Wright-Patterson, I was really in trouble. They dropped me off Instructor Pilot status. Grounded. It was about the worst thing that could ever happen to me. Something like this can wipe you out, in promotions and flying. I was back to flying four hours a month to get flight pay. I was desperate. Then I ran into a guy named Jim Rock who was an instructor in the Aero Club.

This is an Air Force organization with light planes which encourages people in the Air Force or Civil Service to learn to fly at reduced rates. This was another flight plan; I could get my instructor's rating through FAA, if I could get five students through their private pilot's examinations. So, old Jim was back in the air again, hustling students.

That Christmas, 1958, I flew out to San Jose to spend the holiday with the folks. Every time I came into San Jose, I dropped in to see an old friend, "Del" Del Carlo; he ran a

171

photo studio there. He knew all the models in San Jose, and always seemed to have beauties working for him. Del lived across the street from my folks, and he was a part of our lives there. Also, I had made a practice of dating his receptionists.

Del's current receptionist was a very beautiful brunette named Mary Monroe. Del introduced us, and we hit it off. Chuck was with me, and we asked Mary if we could take her home. Mary discovered she had forgotten her key, and just as we were trying to jimmy the back door, her mother walked up. From our first meeting, Mary appeared to be very gathered and organized. She was fairly outspoken; she said what she thought. And when I asked her for a date the next night, she said, "Yes, I'd like to go out with you."

I remember that we had dinner some place over near the studio. For some reason, Mary knocked the peas off her dish. She was so embarrassed, with peas all over the table and her dress. I was kind of amused at her consternation. The day after that we decided to go to the beach. I was not surprised to learn that Mary was a model, that she had gone to the John Robert Powers School. We had a great time down on the beach. I spent two weeks there during the Christmas season, and I saw her almost every night.

Before I left, Mary and I made a little trip down to Monterey. I recall that we had dinner at the Navy Post Graduate School in the old Del Monte Hotel. We just sat there silently eating and looking at each other. I was in love and it was wonderful.

But all too soon, I had to go back to Dayton. I even had to go back commercial that Christmas. I can remember going out to the airport and thinking what a great Christmas vacation that had been. I had found a girl I really loved.

It was not long before I got back on instructor pilot

status. Soon I had another T-33, and I was flying back to the West Coast to see Mary. On this trip we decided that we would get married. We even picked out the place—Ben Lomond Lodge, up in the Redwoods. It was a quaint place with a little brook running through the dining room. They had a little chapel there, and it was really beautiful.

As it happened, on this trip out to see Mary I had been forced to drop my student off enroute at the military hospital in Rapid City, South Dakota, because he had developed serious hypoxia from the high altitude. (There may have been something wrong with his oxygen mask, or maybe his tolerance was low.) After a brief bout with my conscience, I left him in the hospital and flipped on down to California to see Mary. I rationalized that he was in good hands, and actually when I got back to him he was in good shape. But this trip turned out to have a number of consequences.

On Monday morning I got a call from the brass. Dick Goodman braced me about leaving my student in Rapid City. "You are not allowed to do that. Jim, you know you cannot fly an airplane by yourself."

I said, "Dick, I was not aware of that."

He told me that I should have landed with my student when he first showed symptoms, that I was taking a chance on his life. "We are going to have to ground you for thirty days. You can't fly like that," he said.

Grounded! There I was, Mary on the West Coast and me in Dayton, and we were planning to get married. And I was grounded for thirty days. What was I going to do? How was I going to get to the West Coast? I couldn't get out there, so we had to communicate by phone and by letter. But I had the logistics worked out so that I could go from a meeting at Edwards Air Force Base on up to San Jose, get married, and fly Mary back to Dayton. Then I planned to get my hands on a light airplane and fly Mary down to Nassau for a honeymoon. We'll get married in the little chapel in Ben Lomond that we were so struck with. Well, that was my plan.

The day before I was going to fly home, I called from Edwards during **a** recess between meetings. She stunned me. "I don't think I am going to marry you," Mary said. "I just cannot marry you." I talked to her for a little while on the phone, but I couldn't change her mind. So, I had to go back into conference. Man, this just tore at me something fierce: you don't know how my stomach was churning inside of me. Thank God I had only a support role in the conference.

I loved Mary, and I was overwhelmed. I suspected that her father, brothers, and sisters, who are strong Seventh-Day Adventists, had persuaded her that we couldn't get married because of our religious differences.

Why was I always plagued with religious problems? My first marriage was destroyed because of my attitude about my wife's Catholic faith. Now I was facing a similar struggle all over again. I had been upset by Mary's insisting on attending church on Saturday. There were many times when she would come over and spend Saturday with us, and although I went to church with her it bothered me. Saturday had always been a day of activity for me, but it was her custom to take a nap Saturday afternoon. I had been irked by Saturday church and Saturday naps. I guess I was just as much anti-Adventist as I was anti-Catholic.

I didn't change my plans despite this conversation; I went to San Jose. I called Mary when I got there and we talked by telephone. She said again that she did not think that she could marry me. I don't know why I didn't go over and see her. I guess it disturbed me so much that I didn't want to. Mary said later that if I had only come over that night to talk to her, we could probably have resolved the differences and gone ahead with the marriage.

Our wedding plans had included a honeymoon in Florida and Nassau. When I got back to Dayton I decided to take the trip anyway; I went on my honeymoon without my bride. I had this airplane scheduled and I had bought a lot of skin-diving equipment, because Mary liked skin diving. I took all that stuff, and I put it in the airplane. It was foggy. Finally the weather improved and I took off from Wright Field under this low fog. I was fiddling with something in the cockpit, and all of a sudden I looked up and there was a big water tank right in front of me. I whipped to the side and barely missed it. Man, it really woke me up.

I flew on down to Patrick Air Force Base at Cape Canaveral to see an old buddy, Bud Conti. He gave me some sympathy, and I flew on to Miami.

In Miami I filed a flight plan for Nassau. I think I stayed out there a week. I got as black as a native and rode out to remote beaches where I would spend the day diving, swimming, and lying in the sun, trying to forget the world.

Soon after I returned to Dayton, I recovered my status as an instructor pilot and was back in the old routine. Flying cross-country was soothing and comforting for me, and it gave me time to think about my plans for the future—test pilot school.

The truth of the matter is that I had an application in for test pilot school, and my hopes were high because I had the support of Col. Ken Chilstrom. He was my big boss, brought in because the F-108 System Project Office had grown to the point that it required a full colonel. From my point of view they couldn't have picked a better man.

Colonel Chilstrom had test flown the F-80 in the days when they lost a lot of test pilots. He had been in the old test pilot school at Edwards, and he completely understood my desire to go there. He knew how I loved to fly and what my flying capabilities were. Would you believe that at this critical point, with everything going for me, I almost got grounded again?

My boss, Horace Maxey, had a scheduled trip to Los Angeles, and he asked if he could possibly fly with me. So I requested a T-Bird for myself and said that Major Maxey would be going with me. I got a parachute and a helmet for him, and he just flew out with me; it was the greatest thrill of his life. Maxey had never been in a jet before.

Well, Bob Williams found out about it. "Jim," he said, "you didn't tell me that guy was nonrated."

I said, "You never asked me, Bob. He was my boss and he wanted to fly."

Bob really despaired of me, but he didn't ground me that time. He had a lot of instructors working for him—some part-time, some full-time—but I'm sure I gave him more headaches than any of the rest of them.

The paper work was driving me wild. I spent eight hours a day in the office, going to meetings, sorting papers, giving speeches. I got so bored that I was beside myself. But I was a pretty hard worker in the office, and I made some original contributions. So I was getting good effectiveness reports, and in the service your future depends on these reports.

After a couple of months I got a letter from Mary, which kind of surprised me. I thought this thing was wiped clean. She said that she had left San Jose and was living with her sister up in Snoqualmie, east of Seattle, Washington. She said how unhappy she was without me, and that she would like to talk to me. So, I said, why not? I'll do that.

She really looked beautiful, more beautiful than ever.

After that visit I started going out every weekend to be with her. Before long we decided that we should get married. I remember our hiking through the hills holding hands just as we had in the early days of our romance. Now the romance was blossoming again, and she was able to tell me how mixed up she had been. If I had stayed close to her, this would never have happened. Well, we were back in love again and considering our future.

In my own mind I decided that I loved Mary dearly. I vowed that I would marry in good faith—come what may. As far as religion was concerned, I told Mary that when we had children, the girls could go to her church and boys to mine. We set the wedding for Labor Day weekend.

That Friday I flew a student into Seattle, landed on a short, slippery, rough runway, and took off in a cab to meet Mary at the office of a justice of the peace. It was a real sprint —I had a few minutes to spare. Mary was there with the girl friend she was living with. She looked terrific and I remember the justice asking her why she wanted to marry this old man. Afterwards, we went over to the Sorrento Hotel, and that night we had a big prime rib for dinner at the Top of the Town. It was the greatest relief to know that I was finally married to Mary.

The next day was Saturday, so I went to church with Mary and her brother, who lived in Seattle. I put Mary on an afternoon flight to Dayton and kissed her good-bye. Then I rushed back to the Naval Air Station, met my student, took off immediately, and flew back to Dayton. I got to the airport fifteen minutes before Mary landed and took her over to Wright-Patterson in my "new" Karmann Ghia. Then I proudly took Mary to my little basement apartment. She couldn't believe that I was living in this basement. Let me say that Mary didn't feel entirely at home there, so we began looking for another place to live.

We found a little house on a hill surrounded by trees and pasture land. Mary enjoyed cooking and keeping house, and I was proud of this little place we called home. And this was the beginning of our life together—a life that was to be warm, loving, eventful, and sometimes tumultuous.

7

GROUNDED

I WAS ENJOYING the home atmosphere, making it as comfortable for Mary as I could. It was a peaceful existence. I looked forward to coming home in the evenings after work. We had taught ourselves to play chess, and we'd relax by the fireside. Most of our communication was done by touch. We didn't have deep conversations, we just lived quietly together, fitting into this new environment. Those were happy days for both of us, and soon there was an addition to the family.

Mary had a baby girl, and we named her Joy. After four or five days in the hospital, Mary brought Joy home. She took care of the baby and the house by herself. Mary was really great with her. As soon as Joy woke up and started crying, Mary would immediately get to her and take care of her. I never heard much crying from Joy or from any of our other kids.

Joy grew up with the pasture right outside, with cows, sheep, other farm animals all around her. It was a perfect place to raise a child, no doubt about it. In the spring, when it warmed up, I used to take her for piggyback rides down the road. I spent a lot of time with Joy. It was a great experience to have a child of my own.

Meanwhile, the NASA program was under way, and the

first flights were being made into space. To keep up with developments, I began taking Ohio State extension courses in orbital mechanics. When it got warm I would sit out on the porch and study and keep Joy with me. She would play in her high chair or enjoy her little swing that I had hung from the tree. My flying had tapered down to about ten to twenty hours a month, and I had cut out my weekend cross-countries entirely.

In the spring of 1960, I received notification that I had been accepted for test pilot school that summer. This kind of took the pressure off flying and off me, because I didn't want to spend too much time in the air; I wanted to spend all the time I could with Mary and the baby. I guess Mary asked me why I wanted to be a test pilot. I think she sometimes resented my love of flying a little bit, for a love of flying is almost like the love of a woman. Sometimes you can become obsessed with it.

Although Mary may have felt that she had to compete with my flying, she was solidly behind my military career. She had deep values and she was never overwhelmed by rank—but she had a quiet pride in my accomplishments. I am grateful that she didn't drive or push me; she let me do the scrambling. There was never any pressure from home.

We had a month off before test pilot school began, so Mary, Joy and I packed into the Karmann Ghia and had a wonderful trip. We spent a night in Colorado Springs. Mary loved it. Then we went over to Salt Lake City, where I had a chance to visit my old friends, particularly the Woods. We made the big circle up through Mary's old hometown, College Place, near Walla Walla, then over to Seattle and down to San Jose, where we had a nice visit with both sets of parents. We left Joy with my parents and went off to Monterey for a few days. Then we picked up Joy and were on our way to Edwards Air Force Base.

I had driven through the Mojave Desert many times, but I don't think Mary had ever seen it, and it was quite an experience for her. The temperature must have been 110 to 115 degrees. She said, "Jim, why do you want to come to a place like this?" There was nothing green out there; it was all cactus, a few Joshua trees, barren hills, flat dry lakes.

We moved into a very nice older home on the base at Edwards, and I started school. In my class were Frank Borman and Mike Collins, who later became astronauts. Tom Stafford was an instructor. I felt kind of green, but it turned out that I'd probably had more varied flying experiences than any of my schoolmates. There were about fifteen in our class, which was divided into two sessions. One was on performance and the other on stability and control.

We did have a chance to check out airplanes that we had never flown before, like the T-28 and the F-86A. It was interesting flying, but it wasn't challenging. You are not wringing hot stuff out at test pilot school, you are learning techniques that can be used to test airplanes; you are learning to apply standard procedures to measure performance, stability, and control. You are collecting data, and you have to write a report on every test you fly. The secret is to fly smoothly and precisely. You try to hold the air speed to within one or two knots and the altitude within tight tolerances.

We did some spin tests in a T-33. We had to be careful because they had lost a couple of instructors the year before, when their aircraft had gone into a tumble. It just went sideways and fell out of the sky. Tom Stafford was flying with me one day, and he wanted to show me the approach to that type of maneuver, and I'll be darned if we didn't actually tumble

the airplane. One tumble, but we were quick enough to recover it. We immediately released all pressure on the controls, and it flew out of it by itself. Well-designed airplanes
are inherently safe. If you get in trouble, take your hands off
the controls, and a good plane will probably fly out by itself.

The course work was easy for me because of my
academic work at Michigan, so I didn't study very hard. In
the evenings we'd play chess or romp with Joy. I was playing
handball and squash and working out in the beautiful gym.
Then, that February, Jill was born; Joy was only a year old,
so Mary was pretty busy. She was going into Lancaster to
her church every Saturday morning and she slept all Saturday
afternoon. This was a problem, and we argued over our
religious differences.

As my graduation from test pilot school approached I
decided I would like to stay on at Edwards. As a test pilot, I
felt they kept the best guys there.

Finally, my orders came, but they were curious. I was to
stay at Edwards, all right, but I was to report to the Aeronautical Systems Division of Wright-Patterson, which had an
office at Edwards. What was I going to be? A test pilot working for Wright-Patterson again, but at Edwards. I discovered
that I was going to be working with Lt. Col. Allen Nye, a
reserved gentleman with a doctorate, who had taken my old
friend Horace Maxey's place.

Then I finally got the lowdown that this office was working on a top-secret airplane and I was to be the first test pilot.

This was a great honor. I was briefed—I had the clearance for top secret—I met all the people in the office, and
went back to Edwards. I couldn't tell anyone what I was going to be doing, not even Mary. And she became curious because I got to the point where I was not spending very much
time at Edwards. My cover role was that I was the director

of the test force for the ASG-18 fire-control system and the GAR-9 Missile. I was supposed to be going to Hughes Aircraft to work on the Missile and Fire Control, but in reality I was spending most of my time at Lockheed in Burbank.

This was one of the most exciting prospects of my entire life. About a month after I graduated from test pilot school, I decided that I must get a little more flying. They had an Aero Club, and they needed instructor pilots. I'd had only four pilots qualified up to this point and I wanted to get a fifth so I could become an instructor. I was working on both the secret project and the cover project, and instructing in light airplanes on weekends.

My fifth student was M. Sgt. Sam Wyman, who was about forty years old, married, with three children. Sam was a friendly person, a little on the heavy side, a little slow and set in his ways, but I liked him. He was in charge of the photo lab at the base but had always wanted to fly; apparently he had been in flying training in World War II. He was nervous and tended to overreact. He had never had a chance to solo. We flew Saturday and Sunday in the very early morning, over a dry lake bed out there in the Mojave.

The amazing thing about Sam was that as he approached solo he got worse and worse. I told him, "Look, I am not going to hurry you. I'll wait until you are ready if it takes fifteen or twenty hours—just take your time and try to feel a part of the airplane." One Saturday Sam did well in the beginning of the period, but he became worse and his landings were not satisfactory for solo.

The next day, Sunday, we drove out again, I was thinking that if I could just calm him down he might be able to solo that morning. His first landing was pretty horrible. As I recall, he landed tail wheel first and then he bounced and did not make a very satisfactory recovery. After the fourth or fifth attempt, we were taking off with Sam at the controls in the front seat. I was in the back seat. He pulled the plane up too abruptly, turned it too tightly onto the crosswind, and we went into an uncontrollable flat spin.

I never knew whether Sam overcontrolled or froze on the controls so I couldn't overcontrol to recover. We crashed in the desert. Sam threw his arm up to cover his face, but his head went right into the front panel and caved it in. I hit the back of the front seat with my head turned sideways. I was wearing tennis shoes, and when we hit, the front seat collapsed on my feet, particularly my right foot. The seat came down right above my ankle, giving me a compound fracture, with the bones sticking out through the flesh. I had two broken legs, a broken jaw, and a head injury. I was pinned in the wreckage. Fortunately the plane did not catch on fire.

The tower just happened to see us crash. They sent the fire truck and the ambulance from the main base of Edwards down to us; then they pulled us out of the wreckage. I had cuts and bruises all over my legs and arms. We both had severe concussions and amnesia. My memory was wiped out for the twenty-four hours before the accident, so I had to piece the accident back together. Sam's memory was wiped for the five years before the accident, including his relationship with his wife and the memory of his daughter, who was two to three years old. They thought at first that the brain injury was fatal and gave him the last rites. But he survived.

They knew I would survive, but they didn't know what my mental condition would be. They considered amputating my right foot because the circulation was so badly impaired. They took us to March Air Force Base and put us into intensive care. The first person I remember seeing was my dad, who had come down on the bus.

GROUNDED

Mary had been with me earlier, because she came to the airplane to see me off when they made the emergency flight to March. She said that I was wild with pain, that they kept trying to strap me down on a stretcher and I kept taking the seat belt off. I guess I was still trying to get out of the airplane.

It didn't take Mary long to get mobilized and deal with this situation. She came down with the two girls and moved into a trailer just outside of March Air Force Base, so she could be close to me. I recall that when she first brought the girls in, Joy kind of rejected me. She couldn't believe that was her father wrapped up in bandages with a dazed look in his eyes. At that point, my mind was still not functioning properly.

Mary came to the hospital every day, and she gave the nurses a hard time for not taking better care of me. She washed me, and she loved me, and she constantly encouraged me to recover; she really made a difference. A test pilot really thinks it's going to happen to somebody else, not to him. If you didn't think this way you wouldn't be able to fly. You would be too concerned, too cautious, to be a good pilot. But on the remote chance that it does happen, particularly in a testing situation, you assume it will wipe you out. You never picture yourself completely incapacitated. Here I was in this hospital bed wondering if I would ever walk again, let alone fly.

Frank Borman came to see me, as well as my boss, Colonel McIntire, who was in charge of the Aeronautical Systems Office at Edwards. Of course, he didn't know what my secret mission was. I was the only person at Edwards who knew what my true mission was, and after about three or four weeks, after my mind had straightened out, I wrote to my group in Dayton and told them what had happened, that they had better get themselves a new test pilot. This was a hard letter to write.

I had been offered the greatest opportunity a test pilot could ever ask for. I was to have been the first and only test pilot for what became the YF-12A. This would be the highest-flying and the fastest-flying airplane that had ever been built, and flying it would have to be the pinnacle of any pilot's career. It was going to be a Mach 3 airplane (2,000 miles per hour). My career looked finished.

When I was fully conscious and able to think for myself, one of the first questions I asked was, "Lord, why did you let this happen to me? Why did you raise me up so high, if you were going to let me fall so low? Why, Lord? Why?" And as I lay there in that hospital bed, I prayed harder than I ever had before. I prayed for understanding, for recovery. During this dreadful, desolate time, I got some insight. I realized that I was rushing through life too fast; that I was ignoring the daily blessings that were mine. God began to answer the prayers for recovery.

Back there in the hospital I really had an experience. There was this tremendous pain, and the total helplessness of not being able to get out of bed to go to the bathroom. I remember many nights that I couldn't get the nurse, and I would actually urinate in the bed and lie there for hours. It was the most degrading situation. But soon I got to the point of adapting—living with the fact that my legs were immobile, my jaw was wired up, I couldn't eat, I had to suck things up with a straw. Then, after the tremendous pain went away, it got to the point where I enjoyed it. I look back on 1961 as being one of the best summers I ever had. The Lord brought healing, but I would never have made it without Mary.

Mary would come and we had a chance to talk, and she

was the greatest nurse I could ever ask for. She'd shave me, scrub me up, and wash me all over. She enjoyed doing it, and I enjoyed it. Mary and I became much closer as a result of that experience. I began looking forward to getting out, and the fact that she and the girls were right off the base meant everything to me. Before I could be released, I had to manage crutches. I had never used crutches before with two legs in casts, but I got to be pretty good at it. It's like walking on stilts with your arms instead of your legs.

Finally, Dr. Forrest agreed that I could live with Mary out in the trailer, if I came in every day for a checkup. We lived in the trailer for two or three weeks, and I enjoyed our new little baby, Jill. I'd change her diapers or give her a bath, all the time lurching around on crutches. Jill and Joy got a lot more attention that summer than they might have under normal circumstances. The experience really brought our family closer together.

Because my jaw was still laced up, the doctors gave Mary a pair of scissors for an emergency. They knew if I ever got sick, I was likely to choke to death. They permitted me to return to Edwards, with Mary doing the driving. Both my legs were still in casts. My arms and shoulders were in great shape because of all the exercises. But, my lower body was still immobile. My legs were shrinking because of lack of use, but two or three months after the accident there was not much pain in them. Still, the question in my mind was, What are those legs going to look like when they come out of the casts?

We took the cast off the left leg first. It was just a hairline fracture, but the leg was shriveled up. I started doing exercises with an iron boot, and leg lifts to strengthen it. And I started to drive. I mastered the car as a one-legged driver, and I went down to the office on the base and took over my job as test director of the advanced missile. This was the cover job, but it was a real job. We were installing the fire-control system, and the missiles were being built down in Culver City.

When I got in touch with my secret job as test pilot of the YF-12A, I got word back that the program had slipped a little bit and they still wanted me to be the test pilot. I couldn't believe it, but I was being presented with a hope. I couldn't talk to them by phone; I had to write. I was just getting to the point in my recovery when I was strong enough to drive to Hughes Aircraft in Culver City—at least I was able to convince everybody that this was where I was going. Secretly, I went to Lockheed at Burbank.

The guys at Burbank were surprised to see me back. There I was, the Air Force test pilot, coming in on crutches, with a cast on my leg. It amused them, but I think it did something for them to see me hobble out of the car full of confidence and ready to go. Well, this really spurred me on. Soon I was spending two days a week there.

At Lockheed I worked with Kelly Johnson, who was in charge of the new airplane. Kelly had been a chief designer for years; he built the U-2, and Gary Powers, who was shot down over Russia flying the U-2, was working down there as a kind of consultant. Kelly had a big office in a building that he called the Skunkworks. To get anywhere near the building you had to go through guards and have special badges. There the plane, A-11, was being built in complete secrecy by pro-duction-line workers who had security clearance. This big monstrous airplane, which is over 100 feet long and weighs 150,000 pounds, was being handmade, almost entirely of titanium, and its existence had to be kept a secret from the Russians. Just to get the titanium for the airplane and try to keep it a secret in our country is a test of national security.

GROUNDED

Three and a half months after the accident they cut the cast off the right leg—it looked awful. It was shrunken up and covered with nasty scars. There was an open wound on the back of the leg that had been rubbed raw by the cast. Mary was able to accept the horrid-looking thing, but I think it bothered the girls. And this didn't do anything for my morale.

I had been putting weight on the leg, and now I had to start flexing it. The knee muscles were so weak that I went over to the gym every day and took whirlpool treatments. Soon I could really start exercising hard, even jogging a little bit.

Five months after the accident I thought I was ready to fly again. Then I got the bad news: the Air Force has a medical policy that a guy who has had a concussion has to be grounded for at least a year. This was hard news to take. The Flight Surgeon held this information back from me until I had gone through the most critical stages of recovery. I began to think about a career in law, and I actually began taking a La Salle correspondence course.

Mary was fascinated, and we kind of worked on the course together. Within six months I managed to get through a year of law. Now, I had the job as test director of the advanced missile at Edwards and my responsibilities at Culver City and Burbank, and I was studying law. But I was not flying.

Everything was moving ahead with the YF-12A, which we called the A-11, and we were already laying the groundwork for the facilities we would need at Edwards when we brought the plane out in the open. Some top brass at Edwards had to be cleared to expedite this, so they cleared Gen. Irving L. Branch, the Center Commander, and I took him over to

Burbank to see Kelly Johnson. After that, if I had problems with people on the base, I would let General Branch know, and he would contact them. "You work with Captain Irwin," he would tell them. "He knows what he is doing. Do not question him. Just do what he asks you to do." It was great. I almost felt like I was running the base.

In the spring I ran into my old friend Buck Buchanan. He had graduated from the first Aerospace Research Pilot Course at Edwards. Buck, Frank Borman, Jim McDivitt, Tom McElmurry, and one civilian were actually in the first group in the space school. They organized the curriculum as they went along. When Buck finished the course, he took over as Commandant of the school.

Buck knew that I was not on flying status at this time, but he put me in the front seat of a T-28. I remember coming in very fast over Culver City and beaking. Buck said, "Man, you sure like to fly the plane fast." He was right, but I don't think he realized that this was the first time I had flown since the accident. This convinced me that physically I was ready to go again. All I had to do was convince the Air Force. I knew the Air Force was wondering whether my brain was back to normal. This could be resolved only by an electro-encephalogram, or EEG.

Just after I graduated from test pilot school, I was sent down to the School of Aviation Medicine at Brooks Air Force Base for five days of exhaustive tests, physical and mental. They wanted to establish base-line physicals that they could use for astronauts. When they ran me through, they came up with what they call a borderline EEG. Good luck for me that the Air Force did not use the EEG in selecting pilot candidates. But that was not the point.

Providentially the Air Force had this EEG base line on

me, and they knew what my typical brainwave pattern was. If I had not had this control test earlier, they would have been certain that the borderline EEG was a result of the accident, and I never would have flown again. As it was, there were conflicting opinions, and ultimately they sent me back to the School of Aviation Medicine at Brooks, where I went through the whole five days of tests again. They determined conclusively that the new EEG was the same as the one I had recorded after graduation from test pilot school.

I also have a history of high blood pressure, but the Air Force had determined that it would come down to normal once I got accustomed to the cuff.

So finally, after fourteen months, the Air Force put me back on flying status, and my prayers were answered. Life had real point again, and there was a true possibility of test flying the A-11, which was still a dream.

By late 1962, I was doing a lot of flying, and I was intently interested in getting into the Astronaut Corps. With Buck Buchanan's help, I got an assignment to the Aerospace Research Pilot School, which had graduated a number of pilots who became astronauts. I had one of my more interesting flying experiences in executing what they call a "zoom" maneuver in Space School.

We did a lot of high altitude maneuvers to get us accustomed to zero G's as we went over the top of a steep climb, and to give us experience with space suits or full pressure suits. One of my zooms had unexpected results. I was flying a Lockheed F-104 Starfighter with a full pressure suit on, and I accelerated to about Mach 2 (1200 mph), then began my pullup to maximum altitude. I was in a very steep, approximately 40-degree, climb, and at about 70,000 feet the engine was shut down and my suit automatically inflated—I was concentrating on eking out the last bit of altitude. Here I was going through 150 knots, 100, 50, and then the climb stopped at about 85,000 feet. One of our instructors, Bob McIntosh, had told us that it was impossible to spin the 104, even with the stick all the way aft. My aircraft went into a flat spin at 90,000 feet.

My first thought was, wait until I tell Bob about this. I had 40 seconds at zero G's and then I began plummeting down. My air speed hovered at zero so that the antispin controls were ineffective. Tommy Bell was flying chase plane, and I remember him shouting to me: "You're in a spin! You're in a spin!" I was too busy to answer. As the plane dropped through 50,000 feet I had 50 knots air speed; at 35,000 feet I was successful in starting the engine as the spin continued. Tearing through my mind were thoughts of my previous flat spin, and the warning to eject if the aircraft was not under control at 20,000 feet over the Mojave.

At 25,000 feet, the rudder and aileron were fully applied in the right direction, and my air speed was up to 150 knots. Theoretically I was a dead man with too little time to eject. I had no aerodynamics. I still could not pull out; the nose would not go down. A flash of inspiration—I put down the takeoff flaps. It did the trick. At 10,000 feet over the desert I nosed into a dive. I pulled it out about 3,000 feet from the hot sand, and headed home. The radar plot of my zoom showed a steep climb and then a vertical drop from 90,000 feet.

I graduated from Space School in 1963, and with this credential I was encouraged to try to convince the Space Administration that Jim Irwin should be an astronaut.

I got as far as the final selection in Houston, but I was turned down. They don't tell you why, but I am sure it was because of that recent accident on my record. That year my son, Jimmy, was born, and Mary had her hands full, particularly since I was enormously absorbed in flying again and was still leading a double professional life. My ambition was heightened by the challenge of becoming an astronaut—it is amazing how soon I took flying for granted, even the opportunity to fly the YF-12A. This interceptor was exceeding everybody's expectations and was definitely going to be the challenger for the existing world's altitude and speed records.

Jan was born September 30, 1964. We now had four children, and I was the hot test pilot again. The YF-12A did ultimately break the altitude and speed records that were held by the Russians; and, through a carefully executed plan, we broke them both in the same day.

In 1964, I tried again for the Astronaut Corps, but this time Houston was looking for young scientists with doctorates, so I was turned down again. This was a real blow because I was fast approaching the age limit, and it looked to me as though any chance I had was fast diminishing. On Thursday, September 30, of that year, the YF-12A was unveiled before the press and the public at Edwards Air Force Base, and it was tremendously impressive in fly-bys and static displays. Col. A. K. McDonald of the Air Force Defense Command told the press that the interceptor was capable of sustaining flight at Mach 3 speeds and at altitudes in excess of 70,000 feet.

While at Edwards, I made many trips to Colorado Springs to brief the Air Defense Command. They had been vitally interested in all of my work on missiles, fire-control systems and interceptors. It looked as if my experience uniquely qualified me to be in charge of their YF-12A program. I would be doing essentially the same work that I had been doing at Edwards. I knew the total system, I had even flown the plane, and I knew the Air Defense Command. When Col. McDonald, who had this job, was transferred to Japan, it became a foregone conclusion that I would replace him.

This was a staff job, and obviously I wouldn't be doing much flying. I wouldn't be flying the YF-12A. But Mary and I had always wanted to live in Colorado Springs, and everything seemed to fit together.

It was certainly strange for me to request a desk job. But I had experienced so many frustrations at Edwards I was ready for a change. And Mary was never completely happy in the desert, never happy about all that flying. I thought we'd move to Colorado Springs, to a beautiful area and a lovely home, and just enjoy each other and the children.

We drove over from Edwards with the family. We bought a very modern, Japanese-style house in western Colorado Springs and I got everything squared away, even painted the outside of the house, before I went to work. Then I reported to Colonel Hammett, my immediate boss. The hierarchy was Hammett, Colonel Earle, and General Preuss, who reported to General Thatcher, who was Air Defense Command Commander. My immediate boss was not even cleared on the program, but I could talk to Colonel Earle and General Preuss freely. I had three men working for me—Maj. Bill Thurman, Capt. Larry Tibbits, and a Maj. Joe Manning. Here I was, a red-hot test pilot out of Edwards. I had flown the F-12, and a lot of people looked up to me and valued my judgment.

We had to assemble a team that could put together a comprehensive package of the cost and effect of introducing the F-12 to the inventory of the Air Defense Command. In addition to the plane itself, there was also the cost of physical changes on the base, fuel storage, new handling equipment, and new training for pilots and backup personnel. We put a comprehensive package together—briefed the Air Defense Command Headquarters and then briefed it on up to Washington. We had the problem that we always have: people do not believe that there is a viable bomber threat.

After I had been there for six months, it began to look very doubtful that we would ever be able to justify the F-12 to the Air Force. It turned out that the big cost was not purchase but upkeep, the operational cost. It burned a lot of expensive fuel. And yet the command needed airplanes, so we had to build a fallback position. I had to go around the country studying new fire-control systems, missiles and aircraft. We selected the F-4 as our fallback choice for the new interceptor.

We had a little health club with excellent gym facilities right in the building. When I was in Colorado Springs, I worked out for an hour every day instead of eating lunch. I was feeling good; I thought I was getting in better shape, and doing a good job. In the early spring of 1966, I heard they were looking for astronauts again. I figured I might as well try. Why not? The call came out in February, and I was nearly thirty-six—in March I would be at the age limit. For the fifth time, I went down to the School of Aviation Medicine at Brooks.

What really helped me this time was that I had Colonel Earle, General Preuss, and General Thatcher on my side. Colonel Earle thought I had real leadership capabilities, and he really went to bat for me. I found out later that he had contacted every Air Force general he had ever known, writing personal letters to all of them. He had the wisdom to set it up so that the Air Force would give me a strong endorsement as a candidate. And of all of the Air Force people, I think I had the best qualifications. So I went down there for the physical and for the interviews.

Not long after that, Deke Slayton called me at the office. "Jim, would you like to come down to Houston?" "I'm ready, Deke," I said. "When do you want me?"

I had to report the first of May.

8

ASTRONAUT TRAINING

MARY HAD mixed emotions. There was our new little house
that we had invested so much in and had lived in for only
nine months. So many people in the neighborhood had taken
us under their wing. They loved us, and we loved them.
Almost every weekend, we'd drive up to our favorite rock in
Cheyenne Canyon, a five-minute drive. Once you get there
you are completely remote. The kids and I would go swimming in that little creek. Occasionally, we would go over to
the Garden of the Gods—there were so many beautiful spots,
all within a short distance. But I must confess that the chance
of becoming an astronaut, the opportunity for fame, and the
possibility of making more money were also attractive.

I had heard that astronauts got something like $15,000
extra a year.

When I was down in Houston for the interviews, I was
invited out to Dave Scott's magnificent home. Dave and I had
known each other well at Space School. I knew he and the
other astronauts had contracts with Time-Life and World
Book. It seemed like the greatest opportunity you could ever
ask for.

When the news came that I would be an astronaut, we
had a press conference in Colorado Springs. I was the first

Air Defense Command man to be selected, and they made a lot of it.

I drove down to Houston in my little Karmann Ghia and reported in at NASA May 10. The only place they had for me to live was the BOQ out at Ellington Field, old wooden barracks, not quite as bad as Hondo, but worse than Reese. I had a very small living room, a bedroom, and a bathroom— and a refrigerator. Fortunately, in just a few weeks I had checked out in the T-33 and was building up time in the T-38; then I had a plane for the weekend. So my normal pattern was to stay in Houston for the week and fly to Colorado Springs for the weekend.

Since we started right out with orbital mechanics, astronomy, and other difficult subjects, it was just as well that I didn't have the distraction of the family at first. There were a lot of sharp guys there. They had picked nineteen, which was providential; any fewer and I wouldn't have had a chance. NASA was obviously planning to support a much more ambitious moon program than we had any reason to anticipate. We had men from the Air Force, Marines, and Navy, and they seemed to have been looking for people with test experience. Certainly there was an emphasis on engineering, too. We had one Ph.D., Ed Mitchell, who had his degree in aeronautical engineering from MIT. A Navy commander, Ed was the senior man and he became our spokesman. We adopted the name of The Original Nineteen, spoofing The Original Seven a bit.

I became interested in Bay Colony out on Galveston Bay. I thought Mary and the kids would enjoy fishing, crabbing, and sailing. I made arrangements with a developer, John Sheffield, to rent one of his apartments in a new building right out on the bay. But we couldn't find anybody to take the house in Colorado Springs until September, so we struggled through the summer with me commuting and waiting.

At the last minute, Mary rented our house and I drove the family down to Houston in the camper. As we traveled south, it got hotter in the camper, and Houston was just sweltering. Everybody was complaining. We drove over to the apartment, and the children asked, "Daddy, are we going to live here?" I said, "Sure, this is the best Daddy could find."

As a matter of fact, it was rather small. It was a duplex, about as wide as a motel room. There were three bedrooms upstairs; living room, dining room and small efficiency-type kitchen were downstairs. I thought it was great—no yard responsibilities.

My routine during the day was the same as it was before the family arrived, but now I went home at night. During the lunch hour I would work out in the excellent gym facilities—everybody wanted to get in shape, so there were a lot of guys to play handball and squash with. Behind the gym, out under the trees, there was a little quarter-mile track where I could run; I even had a trainer, Joe Garino. It turned out that Joe and I had gone to school together back in Orlando at Memorial Junior High School. I got my yearbook out, and there was Joe's name and his picture. He had been a great athlete and wheeler-dealer. Joe had been the trainer of the original seven astronauts, but he had also done a lot of other things for the guys. He knew the manufacturers of all the athletic equipment, so all you had to do was tell Joe what you wanted and he would get it for you free or at cost. He was just a great guy.

Mary couldn't find a Seventh-Day Adventist church, and when she went to the Presbyterian church with me, it turned her off. Very fortunately, a Methodist minister, Gus

Browning, came by and visited us, so we tried Gus's church and really enjoyed it. He was a down-to-earth man who could preach a good sermon, and he had a good program for the children. We were the only astronauts out there. Maybe we reveled in being a little unique in this particular church. At this point, it seemed as if there was hope for all of us to get together in one church.

I took Mary over to Bay Colony on Galveston Bay right after she got there. "Isn't this pretty?" I asked. "Wouldn't you like to live here on the Bay?"

Mary looked out (maybe it was overcast) and said, "The water is all brown. What's pretty about it? I don't see any hills, or anything. This has got to be the worst-smelling place anywhere. It doesn't appeal to me at all."

Frankly, I already had an option on a big piece of property and the rough plans for a two-story beach house with a deck overlooking the bay. This fell through right then and there. I figured we'd just stay in the apartment and let Mary make the decision.

The process began. One Sunday afternoon we were driving through a community called Shady River, and we stopped and I introduced myself to this fellow standing in the driveway of a handsome adobe hacienda with a tile roof. He turned out to be Jim Gavlik, the architect who had designed the house. He took us through the property and I was very much impressed with some of the features. We decided right then and there that we would get Jim to design a house for us.

Through the process of elimination, we picked Nassau Bay as the neighborhood; we could get a fairly nice lot there, and they gave us $2,000 off because I was an astronaut. We ended up paying $5,000 for the lot. We were right on the corner of this cul-de-sac where Ed Mitchell and Don Lind, both astronauts, had houses. Jim Gavlik designed a house with a floor plan that was almost a copy of the layout of our Colorado Springs house. But instead of a fireplace in the middle of the living area, we created an art gallery for Mary's paintings. The base was a planter area that could be filled with growing things that could be seen from almost every room of the house. We planned a circle of little gardens around the house so that each large window would look out on a garden, and the property was to lie behind an eight-foot fence that would protect all the windows from the street and give us the privacy we wanted.

The first Christmas we decided to get another car. The camper didn't have air conditioning, and it wasn't going to hack it. A guy named Jim Rathman, an Indy 500 racer who ran a Cadillac-Chevrolet place in Melbourne, Florida, was a great friend of the astronauts. I drove the camper down to Melbourne and left it on his used car lot, and picked up a new Chevrolet station wagon at about cost. Then we drove down to Key West for Christmas. This gave me a chance to take the family on a tour of Cape Kennedy. We all got on top of the Vertical Assembly Building, and from the flat roof of that monstrous structure we could look out all over that part of Florida.

Back in Houston we were in the midst of the Gemini era—or, more accurately, we were nearing completion of the Gemini program—and we could see Apollo coming up. My group was not studying any of the Gemini systems. We were working with the Apollo systems, but we were still using the Gemini docking simulator. We would fly this Gemini vehicle and dock with the Agena in this huge room, 200 feet long and 100 feet high, where the simulator was suspended on tracks, so we could go up and down, side to side, and fore and aft. We could physically translate docking with another spacecraft.

We didn't have a lot of contact with the senior astronauts, but they did bring us into the outside contracts soon after we got to Houston. Al Shepard briefed us on the provisions of the Time, Inc., and World Book contracts. Originally, Field Enterprises or World Book and Time, Inc., put up $500,000 a year to be divided among the guys, so when we got there everyone was getting about $15,000 each. Then Field Enterprises dropped out, and this reduced the pot to $200,000 per year. The forty of us wound up getting about $5,000 a year apiece—a far cry from the bonanza we had heard all the stories about. The contract had the effect of giving Time-Life the first shot at us and made our families available to them. Actually Mary and I had less money left over than we did in Colorado Springs.

Some of the early astronauts did extremely well. They each must have received from $50,000 to $80,000 a year, so they had enough money to make some good investments. They built motels and invested in all sort of things and got set financially.

Our ground classes were taught by visiting experts from Grumman, where they were building the Lunar Module, and from North American, which had the contract for the Command and Service Module. As we got into the courses we had to decide which vehicle would be our specialty. I felt partial to the Lunar Module—it was much simpler and was more like a fighter plane in its design. But the real attraction for me was that if you specialized in the Lunar Module you increased your chance of getting on the surface of the moon.

The guys divided just about evenly, and I chose the Lunar Module. No one flew as a commander on his first mission, and at that point my group was just hoping for one mission.

Grumman was just about to launch Lunar Excursion

Module 1, and they were already working on LEM 2 and 3. They were new at this, whereas North American had had a year's experience checking out the Command and Service Module. Initially, I had my reservations about Grumman's capabilities and about their people. It was my first exposure to New Yorkers. They were cold and abrupt in the way they talked and acted. I wondered about their dedication to their jobs, because they didn't seem to care about anything. As I got to know these people better, however, I realized it was a matter of style. At about this time Neil Armstrong had decided that I would have prime responsibility on Lunar Test Article (LTA) 8 as crew commander, and I would put this vehicle through a series of thermal vacuum tests.

We were pretesting systems that were being incorporated in later models that were for flight. In order to approximate the conditions the LEM would encounter, we tested the vehicle in what amounted to a gigantic thermos bottle that could reproduce the pressure and temperature of space. Lt. Comdr. John Bull was assigned to work for me on the project. This entire experiment was in many ways the most challenging and the most rewarding experience I ever had. In a personal way, it was almost more rewarding than the trip to the moon.

The research centered in Bethpage on Long Island, and the vehicle was built there. It was tested in the factory and finally flown down to Houston, where it was put in the thermal vacuum chamber. One reason this project was so satisfying is that I was in charge, I was the commander. It was my program.

The disastrous Apollo 1 fire on January 27, during a routine ground test, killed the prime crew. Because of this tragedy, NASA was forced to undertake exhaustive reevaluation of everything used in the program in terms of its flammability. They even changed the environment at launch from 100 percent oxygen to a mixture of nitrogen and oxygen, which is less flammable. We went to Beta cloth for space-suit material, and I had the distinction of running the first tests in that suit. This new emphasis put additional burden on the tests that we were doing with the LTA-8 to make sure it was as safe as possible.

Before we did any thermal work, we had to run through some emergency egress tests. Just as we were getting involved in this, John Bull started developing curious symptoms. We played handball almost every day, and I could see that John was losing his stamina. Occasionally, he would have coughing spells. The doctors discovered he had a progressive condition they called aspirin asthma—if he took an aspirin, it would constrict his lungs and he couldn't breathe normally.

NASA sent John to Stanford, where he got his doctorate; when Deke decided he couldn't spare another astronaut to put on this test, we borrowed Gerry Gibbons from Grumman. He was almost the same size as John, and since Gerry was a consulting pilot working on the LTA-8, he had been following all this stuff and had the procedures down. All he had to do was learn the nomenclature of the equipment and follow instructions. We made a smooth transition.

Running these tests was almost like getting ready for a mission. We'd go into the suit room, where they would put the sensors on us and suit us up, and then we'd prebreathe for three hours to get the nitrogen out of our systems. Then they would pump the air out of the airlock, and we would get into the vehicle. After eight hours in the vehicle, we'd reverse the whole process, working back through the three umbilicals. You have to be very painstaking, because if you miss one step you've had it. You could hurt yourself and ruin the vehicle and zilch the timetable of the Apollo program.

I had a startling experience during one of the emergency egress tests, requiring us to get out of the chamber at high speed. I was to put on an oxygen mask, crawl out through the front hatch onto the porch, and walk down the steps and out of the chamber. I started the exercise by putting the oxygen mask on and turning the valve to what I thought was *on*. As I was getting out of the spacecraft, my vision began to blur, I started feeling dizzy, and the last thing I remember is trying to stand up on the porch. The next thing I knew I was in the man lock with the guys holding me. Just as I had reached the edge of the platform, I had passed out. Fortunately, the chamber rescue crew had come in and were standing at the bottom of the ladder, and they caught me as I fell down the steps. Otherwise, I would have clobbered myself, wiped out the test, and maybe derailed the program.

After an exhaustive examination, they made a simple discovery. I had never operated that oxygen valve before, and had not turned it completely into the detent position; it had rotated back into the *off* position, cutting off my oxygen, and I had passed out.

In October, our house was completed and the family moved in. There was plenty of work involved, such as laying out and arranging the patios. I actually put down redwood flooring in a checkerboard pattern and Mary planted things between the squares. Jimmy had his bedroom and the three girls had two bedrooms that adjoined.

We had a little study for me and a little study for Mary, with a skylight so she could get that clear, natural light she needs. Mary had been painting since we were at Edwards.

I actually got her started. I was in this little gallery near Edwards and saw this painting done with a palette knife that I liked. I was intrigued with the technique and wanted the painting. Yet they wanted $100 to $150 for it. I said, "This is ridiculous, I can paint that." So I went back to the base and created something that looked like nothing you have ever seen. Mary got interested and became involved; she even took a few lessons. She has natural talent, and she has been painting ever since. I really appreciate her paintings, and finally we had a place to hang them. Now it is interesting to me that Joy and Jill are taking up painting.

Mary has always been allergic to smoke and dust, so we had a built-in vacuum cleaner. You just plug it into the wall and it automatically turns on its motor and sucks the dust and dirt into a central receiving place. We even put in an electronic air filter that purifies the air and keeps the dust out of the atmosphere. So, we had a very comfortable home.

Then, suddenly our lives were disrupted. We got word that Mary's twenty-two-year-old brother Clair, a young bridegroom, had been killed in a tragic hiking accident in the Santa Cruz Mountains in California. He had been hiking with his dog in the mountains west of San Jose, and he had evidently walked too close to the edge of a dropoff, and the edge collapsed. He was killed as he fell into a ravine.

I went out to the service in San Jose with Mary, and I thought she was very brave. I flew back to keep up with the astronaut program. After a few days Mary brought one of her sisters back with her, and she spent a week or so with us. They had a lot to talk about, and then Saturday came up, and they decided to go to church. They found a little Seventh-Day Adventist church up in Pasadena, Texas, about twenty miles north of where we were living, and from then on Mary was back in her old church. Somehow the death of her brother caused her to feel she ought to get closer to the Lord, and the only way she felt she could do that was to go back to her own church.

ASTRONAUT TRAINING

By this time I had heard of the Nassau Bay Baptist Church and had met Bill Rittenhouse, the pastor. He frequently came over to the Space Center with visitors, and I had met him in the astronaut's area.

Now, Mary normally left the house at 9:00 or 9:30 A.M. on Saturday, and she got back about 12:30 or 1:00 P.M. When she was in church I would cut the grass, do yard work. I couldn't do anything with the family because Mary had gone off. She'd come back and fix lunch for the family; then she'd take her nap. This was the period when she decided to become a vegetarian.

About the time I was reaching a peak of agitation and frustration about this, Mary and I had the opportunity to take a vacation together, without the children. This was probably the first time we had ever been able to get away together since all the children were born. We were going to Acapulco, to Las Brisas, a fabulous hotel on a mountaintop overlooking Acapulco Bay. Each room is an individual cottage, and each cottage has its own jeep and own swimming pool, on a different level, and the service is fabulous. We were going to spend a week or two there.

We flew down. The first night we were alone, having dinner and looking out over Acapulco Bay, when the waiter came over with a message for us. I had to call Houston. I lifted up the receiver and there was Dr. Jim McGee, one of our doctors at NASA. He said, "Jim there has been an accident at home. Three of your children are in the hospital. They were over playing on a house that was being built on an adjoining lot and the framing fell over on them." You can imagine our thoughts.

We had just arrived in Acapulco, we had just sat down for dinner, and three out of four of our children were injured.

How seriously? He didn't know the extent of the injuries at that time, but Jimmy had a rather serious concussion and perhaps a skull fracture.

We got on the phone to get the next flight out.

I had been having such a difficult time at home that I had actually considered going down to Mexico alone. I had thought maybe I would enjoy it more down there by myself. I needed to get away from it all. Then, when the accident occurred, I wondered if I were being punished for having had those thoughts. I vowed never to go on vacation again without the children.

We flew back to Houston the next morning and drove to the hospital. Joy and Jill were shaken up, bruised and cut, but they were ready to go home. Jimmy had a hairline skull fracture, however, and he had to stay in the hospital for another couple of days. The kids were really a little frightened to see us; they wondered what our reaction would be since they made us come all the way back from Mexico. They needn't have worried. We were so relieved that we bundled them all up, as soon as Jimmy was out of the hospital, and took them away with us for a week.

The LTA-8 tests were successfully concluded, and we had refined procedures and instituted modification of some of the equipment. I was proud of our test team. They had performed well under the observation of NASA, the Grumman test directors, and a whole staff of people who had been monitoring all the systems.

This project had caught me up so that I was not even thinking about a trip to the moon. I wondered what I could possibly do to equal it. Then I found out. Tom Stafford gave me a call. "Jim," he said, "how would you like to come to work for me?" This was the spring of 1968, and I took the assignment in the summer. I knew that Tom was going to be the crew commander for Apollo 10, so I told him that it would be a great opportunity and I would welcome it. I thought Tom wanted me to be backup for 10. As it turned out, he wanted me to be support crew, the third tier down. Ed Mitchell was going to be backup. This was a blow to my ego.

The Apollo 10 crew were Tom Stafford as Commander, Gene Cernan as Lunar Module Pilot, and John Young as Command Module Pilot. Backup to them were Gordon Cooper, Ed Mitchell, and Stu Roosa. Stu Roosa was a contemporary of mine and Ed's—so there I was supporting my own contemporaries. Joe Engle and I made up the support crew. We were supposed to do the legwork for these guys. The crew can't cover everything, so we try to be their eyes, make sure the equipment is working right. This was something of a step down.

I started spending more time at the Cape because the hardware was in the Vertical Assembly Building. I actually liked to get in and use the hardware, so I would spend hours in the cockpit along with the Grumman technicians, and I became the LEM expert as far as they were concerned. Many of the astronauts were pretty haughty. They wouldn't get in there with the Grumman technicians unless it was a very important test. Of course I was working with Tom Stafford a lot. Tom was a year behind me at the Naval Academy, and this is a bit humbling. He had fewer setbacks. (He was the guy that tumbled the T-Bird with me when he was showing me the maneuver to avoid.) He is quiet, like I am, but very canny and popular.

Tom developed a specialty of understanding rendezvous procedures. Deke recognized this talent, so any time there was a new rendezvous situation, Tom was called in. Tom has command of the docking situation with the Soviets.

About this time Dave Scott approached me and asked if

I would like to be backup with him on Apollo 12. I was delighted. I said I would love it. He said he would have Al Worden and me—so the team was formed. Dave knew all the experience I had had with the Lunar Module, and it was in his interest to get a good Lunar Module man. Dave himself had had a lot of experience on Apollo 9 in the Command Module. Jim McDivitt and Rusty Sweickart were in the Lunar Module and Dave was in the Command Module. They didn't land on the moon, but they did rendezvous. Not only was Dave one of the most competent people in the program, he probably knew most about the Command Module.

We were testing the LM at Grumman and the Command Module at North American in Downey, California. (The Los Angeles area is so fabulous, many guys pick the CM.) Al was working with us; since he was on support for Apollo 9, he knows a great deal about the CM. This was essential because Al would be left by himself while Dave and I were on the surface of the moon. The prime crew for 12 was Pete Conrad as Commander, Dick Gordon as Command Module Pilot, and Alan Bean as LM Pilot. I was backup for Alan Bean.

About this time I began to major in rocks. I had always loved mountains, and rocks really interested me for their beauty. But getting down to studying individual crystals and trying to identify them left me cold. From the beginning of astronaut training we had this ground school of basic instruction in geology, mineralogy, and petrology (the chemistry of rocks). But once I was chosen for Apollo 12 as backup, a geologist was assigned to give us specific training. When you get assigned to an Apollo crew that is to land on the surface of the moon, geology becomes very important.

We made a geological trip about every two months somewhere on the North American continent in areas with characteristics similar to the moon. One of the hazards in this pursuit was Jack Schmidt, who was geologist before he became an astronaut. He acted as one of the instructors, and he would lecture to anyone who would listen at any time. Even if they weren't listening, he would lecture. We'd tell Jack to be quiet and let us hear a real instructor.

There was one terrific geological trip to Brooks Lake in Katmai National Park in Alaska. The most recent volcano in the United States (active in 1937) is nearby, and we were able to study the ash and debris. Big volcanic mountains with snow on them were nearby, and the Army had supplied two banana-type choppers to lift us up to the base of the volcanoes. We stayed at the Brooks Lake Lodge, a fishing camp. The salmon were spawning then, and huge fish were leaping out of the streams all over the place. After geology all day, we did a little fishing.

I was fishing one afternoon, when I noticed a little movement across the stream. There was a huge Kodiak bear. (They're the largest carnivores in the world.) I watched him come up the far bank and I was absolutely fascinated; then he started crossing the stream toward me. He looked at me, and I looked at him. It didn't bother him at all. But I wasn't taking any chances. I took off running, and I ran the whole mile back to camp. I really was glad that I had done a lot of jogging. When I got there, I discovered that I was still clutching my fishing rod.

These trips were great for me. I liked the break, and I enjoyed getting out into the wilds. There was always time to relax after the work was done.

One of the most interesting experiences we ever had was the jungle survival trip. They figured that they could find us anywhere on earth in two or three days, so they were interested in short-term survival. And since it seemed very unlikely we would come down in snow, they concentrated on the sea, the desert, and the jungle. We found some real tropical jungle down in the Panama Canal Zone. And we got a little glimpse of how we were going to live if we splashed down in a rain forest instead of on the briny.

We flew into Albrook Air Force Base for one day of basic instruction on general survival. Our luncheon that day was a memorable gourmet meal. The appetizer was braised boa, our entrees were iguana thermidor, oven-roasted wild boar and baked armadillo. We had heart of palm and bamboo shoots as the salad, and our dessert was tropical fruit compote. We were told that we could find the same menu on survival trek. The next day they took us out in helicopters and dropped us off up on a ridge in the rain forest.

We hiked in about two miles and got down on the edge of a little stream in heavy underbrush with rotted debris under the trees. This was where we were going to live for three days. We were wearing flying suits, long underwear, and tennis shoes. Since it rains every afternoon, we went right to work putting up a lean-to. Sure enough, it started raining before we finished and we got soaking wet. One astronaut was down fishing in the stream, when he looked over his shoulder into the eyes of a big leopard. They took off in opposite directions.

As it turned out we had heart of palm for breakfast, lunch, and dinner. After three days the guides gathered us together and led us out to the Chagris River. We inflated some life rafts that we had and floated down the river about ten miles, until we came to Antonio's Village on the Chagris. What an address. Antonio, the Indian in charge of this little village, had invited us to have lunch with him.

We went into the main hut and sat on the floor with

Antonio and a lot of big iguanas, three or four feet long. I think they had their eyes shut, but they were just lying there with their back legs tied. I guess they were the meal for tomorrow or the next day. We did have roast iguana, but it wasn't nearly as good as the iguana thermidor at survival school. The thing that attracted our attention was the chief's bare-bosomed wives. Our instructions were not to stare at the girls too much. After the banquet they flew us back to the States.

We were notified late in 1969—that is, about a year and a half before the flight—that we would be the prime crew for Apollo 15. We had already trained for approximately a year and a half as backup for 12, and when Apollo 12 was launched we were out of a job until they decided that we were really going to move into the prime position.

At one point I was told by Dave that there was strong pressure from the scientific community to put Jack Schmidt in my position. Since NASA is responsive to the nation's scientists it was touch and go there for a while, but evidently I had the backing of Deke Slayton, Al Shepard, and Dave Scott.

Jack did a very good thing. He had urged NASA to get a distinguished geologist dedicated to moon exploration to work with the guys, not only to train them but to inspire them to think geology. And one of his professors from Cal Tech, Lee Silver, took over as a consultant for NASA to work with Apollo 15—that's us. He was the greatest guy, just a ball of fire. A sandy, red-haired fellow with a red face, he looked like an old cowboy.

We trained in the field just as we would operate on the moon. Dave and I would work as a team, and we'd have a Capsule Communicator, usually Little Joe Allen.

Dave would select the rocks by placing a gnomon in

such a way that it gave an indication of the vertical, and he'd check the chart for color classification, and we'd take a total of five pictures for each rock. I supported Dave, as a rock carrier. We worked our routine out carefully and practiced it over and over. Every month we went to a different geological site. As it turned out, none of them were as interesting as Hadley Rille.

Hadley Rille had to be the most interesting geologically and topographically, because of the rocks themselves and the dramatic site, tucked in among those high mountains. We strongly supported the selection of the rille as a landing site for 15, and we were delighted when this choice was made.

As soon as the site was selected, everything started moving. They made maps of the area, topographical relief forms of the site that could be used for the landing simulator. These forms were placed on the ceiling in the simulator building. About six months after the selection of the flight crew, we had this model in operation so that we could practice landings. A TV camera would move transversely in our downrange direction, and as we got closer to the moon in our simulator, it would give us a view out both windows of the LM. It looked just like we were landing on the moon. And we landed hundreds of times.

As crew we participated in all the tests right up to and including the countdown. We had our space suits, three different suits for our flight: the prime suit, the backup suit, and the training suit. Each suit costs about $60,000. We had to go to Dover, Delaware, to have them fitted, and this took half a day or more. As a backup crewman you have the same $60,000 suits you have as a prime crewman. Actually, you save the country from $60,000 to $120,000 if you move up from backup to prime. As to the space suit, you get the basic first layer fitted and adjusted; then you put on the layer that gives thermal, micrometeorite protection. This is an expensive wardrobe, comparatively, but really the whole trip was pretty reasonable—about $500 million, mostly hardware.

During these months I was trying to devote all my attention to getting ready for the flight, but my anger and hostility about the home situation kept me unsettled. I couldn't just calmly do my job. These strained relations distracted me. Mary felt that the Lord was working through her family tragedy to get her heading in the right direction. She said, "The Lord knew what it was going to take to spin me around." It seemed an unnatural response to me.

I thought the woman should be there to assist the man, help him in his task. None of this helped me at all. I got to the point where I was just facing each day and trying for the best job I could do on that particular day. Then I would wipe it clean and start out again. If you looked at the total of what a man had to learn to get ready for one of those flights, it was almost overwhelming. I needed the right environment at home to help me.

Often when Mary and I are interviewed, or when we are on the same platform, somebody will say, "Well, the credit for your success really goes to the woman who has been by your side all these years." That amuses me, and it amuses her. Over the years Mary has given me support, and I love her, but many times the tensions at home almost caused me to leave the program.

During that last year of training for Apollo 15, there was so much emotional turmoil at home that I couldn't concentrate on the job at hand. Many times I seriously considered going to Deke Slayton and saying, "Deke, I can't do it any more. There is so much turmoil in my private life that I can't concentrate on the job at hand. Just take me off the crew. I don't think I'm going to be ready for the flight."

There were many times when we talked openly, in front of the children, about getting a divorce. It was a terrible mistake. They were really threatened because Mary and I were not getting along. We wouldn't fight—I'd just be silent —but these long periods of quiet would bring the children close to tears. Then the air would clear and we would come back together, and the children would get over it.

About a year before the flight, I was frequently involved in geological trips, survival tests, and visits to manufacturers of the components—I know it was hectic for Mary at home. She had the whole burden of four kids, the house, and everything. When I was home, I was completely absorbed; I was like an automaton. I had been so programmed I was like a robot. I had to eat, drink, sleep, and dream my work—I had to be saturated. I know she understood that it had to be this way before a flight, but it didn't make it any easier.

When I told Mary that I might drop out of the program, she said I should do what I wanted. She distrusted the glory and fame and adulation that are heaped on astronauts. It wasn't part of her life. To begin with, you have to make a real adjustment to be the wife of an Air Force officer. Being the wife of a test pilot is even stranger. And being the wife of an astronaut is strangest of all. Mary would be quite content to be the wife of a farmer. It would please her to be out somewhere in the country, completely by herself, surrounded by animals. She'd like to be with goats and horses and a few sheep and a dog—and me, I guess. I would like this, too. I'm a loner. But unfortunately I also like to fly airplanes, fast airplanes.

Frankly, I do not care for all the fame and publicity, but it is something that naturally happens if you are a test pilot. Mary is a personality somewhat like me; we don't have many close friends. She is a loner too. Mary went to a few of the coffees that the astronauts' wives arranged, but she didn't enjoy them. She is not a coffee-type girl; she has no time for petty gossip or lady talk. She wasn't about to unload her problems.

About seven months before the flight, Mary decided she had had it. She says that one night she promised herself that the next morning she would file for divorce. It seemed to her that neither one of us was willing to give, and that we were destroying each other. She had gone so far as to draw up two lists: one giving reasons why she should stay with me, and the other why she should go. There was a long list of reasons why she should leave, and only two why she should stay. Mary says that the Lord turned her around, and that He pushed her and kicked her all over the place, and she did not leave. As a matter of fact, from then on things slowly began to straighten themselves out for both of us.

The last couple of years before the flight weren't all bad. There were certain advantages in being an astronaut that I really enjoyed. First, General Motors offered us a car on a lease option for just token money. For a couple of years I had a red Corvette. Man, that's living. I was really on top of the world. I guess the highest speed I ever got out of that car was about 120 miles per hour. Then Ford joined the action, so I got a red Mustang convertible. Two new red cars in the garage.

Then the crews started getting cars that looked alike, so they could be recognized as they traveled back and forth from the Cape. Apollo 12 had gold and black. Dave Al and I decided that we would be the All-American team and go red, white and blue. I had red, Dave had blue, and Al got white. As we moved closer to flight, we got later models, and we had them put racing stripes on our cars. I had a red Corvette with a blue and white racing stripe. Mary was very embarrassed to have those two brand-new red cars in the garage.

Six months before the flight, we left Houston and began our final phase of training in Cocoa Beach, Florida. Those three Corvettes—red, white, and blue—were rapidly identified and earned celebrity status in short order. I was a thousand miles away from Mary and the children, but things were going better at home, and I was totally absorbed in the training. I got home on weekends, I called a lot, and Mary and the children came down and spent about a month at the Cape that summer. Several days they came over and watched us work on the sand pile, all suited up, using the tools and practicing the traverses in our simulated Rover. Mary and the kids even had a chance to watch us practice in the Command Module and the Lunar Module simulators. We really had a good time together, and Mary and I were getting along much better. The worst seemed to be over.

The little Lunar Rover that we had at the Cape was a 1 G simulation that General Motors built for us for about $1,000,000. It looked exactly like the real thing. It had rubber tires and electrical drive, and it responded just about like the Rover would on the moon. We just didn't get the impression of speed on the earth that we were to get on the moon. Ten miles per hour on the earth seemed sluggish, but we just flew on the moon. I mean, we were spaceborne. There is no way we could have driven the lunar version on the earth because it was far too fragile; it only weighed 500 pounds. The extremely light frame would support the load on the moon at 1/6 G, but it wouldn't have taken earth shocks at 1 G. I guess our earth version weighed about 1,000 pounds.

We got four Lunar Rovers for $32 million—that's how I figure $8 million for ours. Boeing built them, and Marshall

Flight Center supervised the job and was responsible for engineering and development and production. If it seems expensive, you have to realize that they were designed and built for a whole new world. All sorts of mechanical, thermal, and electrical problems had to be solved. The cooling system had to be rather unusual. They tried to radiate the heat with a radiator, but they couldn't get rid of it all, so they put seven pounds of beeswax in one of the electronics units to absorb the heat. The electronics in the unit generated heat, and we had the solar heat coming in, so we had mirrored surfaces on top of heat-generating devices to radiate it, and we frequently had to brush off these surfaces to help the process of radiation.

Since the temperature range on the moon is 500 degrees, from minus 250 during the lunar night to plus 250 at moon noon, you have to have a fairly innovative vehicle.

I drove Mary and the kids back to Houston after their visit. I didn't know when I'd see them again, but it was a composed good-bye—we'd had good-byes every weekend. Then I went back down to Cocoa Beach for the last stretch. We kept going through the whole drill.

We simulated solutions to all sorts of possible drastic problems. One of the most interesting was to achieve recovery from a faulty Lunar Orbit Insertion Burn, the sort of crisis you'd have to face if the engine shut down prematurely when you were making the burn. We would assume that the burn was a failure, and that we couldn't use the engines in the Command Module. In that case you have thirty minutes to transfer to the Lunar Module and make the burn in the LM or you can't get back to the earth. You are behind the moon, so Houston can't help you. You are on your own; you make the decision. You have to transfer, power up the Lunar Module, go through the checkout procedure, and make that burn in thirty minutes or you are a permanent inhabitant of space. There is food in the Lunar Module to sustain you if you have to retreat to it.

So you make the burn, and when you come around the back side of the moon and are in touch with Houston, you say, "How did it look? Did we do the burn correctly?" The procedure is known as Lunar Orbit Insertion Abort, and I am very happy that it was only simulated.

We practiced all sorts of aborts. They would let us touch down on the surface of the moon; then all of a sudden the vehicle would start tilting over. If it passed 45 degrees we would have had it, so before that happened we would push the button and light off the ascent engine.

We practiced solving problems on reentry. Of course, the nominal for reentry is to let the computer fly it in, and you can anticipate the point at which the computer will roll the vehicle to ensure capture by the atmosphere. If the computer doesn't roll it, the crew takes over and does it manually. You have a couple of seconds.

Toward the last we would spend half a day in the Lunar Module Simulator, half a day in the Command Module Simulator, and then we'd swap off and take half a day practicing geology with the tools and suits. When we were suited up it was almost unbearable in the Florida summer; we'd sweat like mad. You'd take your gloves off and just pour the water out of them. We had cooling packs on our backs, and every hour we would have to take a break and reservice the backpack, put some more dry ice in. They had gallons of Gatorade, and we just guzzled it. All this exhausting work in the extreme heat is one of the things that may have depleted our potassium level.

As crew we had a great deal to say about our schedule. If we thought we needed special new information or review, we scheduled it. We spent many evenings being briefed and going over emergency procedures, and the geologists came in and talked to us a couple of times a week. Deke had the ultimate responsibility for preparing us—he had to be sure we were ready. He was a great manager: managing our education, our training, and our experience. It is not just a matter of formal education, it is a matter of experience, of working with a piece of equipment until you really know it.

There wasn't any conscious effort to prepare us psychologically for the trip, but I think quarantine did this. There was very little outside distraction. You are confined to quarters, the simulator, and you have the opportunity to concentrate on and internalize everything you will need for the flight. We had the option of sleeping in the crew's quarters or out at the beach house. Frequently I would go to the beach house. I enjoyed sleeping out there on the beach just listening to the ocean, completely alone. I'd get up in the early morning and go for a dip or go for a run on the beach. I have always been a sun worshiper. And that was real peace and quiet, lying there on the beach and listening to the breakers and the sea gulls.

I didn't swim too far out in the ocean, just out to where the breakers break. I didn't want to lose it all at that point. I didn't want to be taken by a shark. Frankly, I thought about that—coming back with a leg chopped off and saying, "Where's my backup?"

Mary decided she would not spend the last weekend before the flight with me. Lurton Scott, Dave's wife, had called her, and they had talked about coming out to see us. Lurton had told Mary, "What if you are harboring a bug, and suddenly it erupted during the flight?" The more she thought about it, the more Mary agreed with her. She decided not to come. She said, "Don't ask me to come out. I

221

want to come so badly." I didn't ask her, and she didn't come, and she was very calm about the flight. She said that the Lord had given her complete assurance that she had nothing to worry about.

The day before the flight, Mary was ill. "I was so sick," Mary told me, "I was passing out. I couldn't keep anything on my stomach. I was so grateful I hadn't given in to my inner feelings and come to see you." Well, I was glad too, retrospectively, after I understood some of the creature problems of the flight.

The last time I saw Mary and the family before the flight they were behind glass, and the rest of the family did most of the talking. But I had a sense of her calm, and I knew the children thought I would get back all right. After all, everybody else had.

When we blasted off, I was absolutely dedicated to the challenge of achieving a perfect flight. I was thinking only about the scientific aspect; I didn't have any notion of the spiritual voyage whatsoever. It would have been beyond my wildest imagination to guess that this flight would not satisfy me, that I would come back to earth a different person, bound for a higher flight.

9

HIGH FLIGHT

On THE FLIGHT HOME to earth there were times when I dreaded coming back. I had the greatest feeling in space. When we splashed down, I wanted to keep time from moving so fast. There in the soft, warm waters of the Pacific, I felt a joy at being back on earth, but I knew intuitively that this was the last calm time I was going to have. I was going to have to make speeches, to meet people, and to take part in public occasions. I had never been able to speak extemporaneously, had never been at ease on my feet. I was a quiet man; I never talked very much to anybody. I had this feeling of being in a new life, of being a new man, and I was afraid of it at the same time. Yet I didn't understand what was happening inside of me. I wouldn't face it.

When we had our press conference in Houston on Splashdown plus five, we weren't completely debriefed, nor were we through with our medical examinations. I knew I had not gotten back to normal physically or emotionally. I resented being called on for a response to the flight because I didn't know what had happened to me yet. But at least I did say that I had felt the presence of God in a very profound way on the face of the moon.

We all knew that our crew would belong to NASA Public Relations for three months. The crews of Apollo 11 and Apollo 12 went most of the way around the world as goodwill ambassadors. Since then NASA had started cutting back, but we were going to do a good bit of traveling as soon as Houston was through with us. Our first visits were to New York City, Salt Lake City, and Chicago. I had never been in a parade in my life until that trip to New York. It was an incredible emotional experience.

We flew into La Guardia in a NASA airplane. A police escort took us in to meet Mayor Lindsay somewhere uptown, and then I think we had lunch. We got into a motorcade with the three of us in an open-top car, and we drove right through the center of Manhattan, down Fifth Avenue. It wasn't a full-blown ticker-tape parade, just a motorcade. John Lindsay was sitting on my feet so I wouldn't fall out of the car; another guy was sitting on Al's feet, and another on Dave's. We had been cautioned not to put our hands out of the car but to take our watches off anyway, to keep anybody from snatching them if we forgot.

To look out and see all those people on all sides—up and down as far as you could see—was really moving. They were even leaning out of the windows trying to see us. To look into their faces and see their shining eyes and their expressions made you want to respond to all of them. I was grinning, laughing, waving, and trying to see as many as I could see. It was a beautiful clear fall day, and when we got down to City Hall the Mayor made a speech and presented the medals of the City of New York to us. We each had a chance to respond with a few remarks, and I was last again. We still spoke in the order of our couches in the spacecraft.

I told them how grateful I was that I had had the chance to make the flight, how proud I was to be an American, and how proud I was to be able to put our flag on the moon—to be there for a million or maybe a billion years.

And I tried to share the emotion I had felt during the parade. I said that as I looked into the eyes of the people of New York, I was reminded that truly I made the trip to see the moon for all of them. And I hoped they were lifted up with us and sensed the excitement we felt. I said that really we didn't go for ourselves, that we were bringing back scientific knowledge; I didn't yet realize that I had brought back with me a religious message I had to share. At that time, I had only spoken to one or two churches, and I didn't yet understand that my religious experience was the most important thing in my life. Looking back on it, the significant thing then was that these people felt that we were theirs, that we belonged to them. We were their heroes, but we were also their representatives, and we tried to tell them that what we had done belonged to them.

We went from New York to Salt Lake City, not to Los Angeles or San Francisco, for the simple reason that James Fletcher, a Mormon who was at that time the Director of NASA, had been president of the University of Utah. This was the greatest choice for me, because Salt Lake City was my adopted hometown.

Man, they really laid on a schedule. Helicopters were waiting to take Dave to Ogden and to take Al and me to Provo, where we were to speak at Brigham Young University. I had an opportunity to go out to the University of Utah, which isn't far from my old high school, East High. They had the Glee Club from my high school to sing their own rendition of "First Steps on the Moon," and I was really overwhelmed.

We went down to a Rotary meeting in downtown Salt Lake City, and my old physics teacher, Mr. D. E. Powelson, was there sitting right next to me at the luncheon. I remembered him right away. I told him he made me love science, and this changed my life. Of course there were many other friends at that luncheon, too, including the Woods family. Bill Woods had played an important role in my life from the moment he got me the job at Makoff's.

We had a parade down Main Street, and a good crowd turned out. I think the people of Salt Lake City appreciated Jim Fletcher's bringing the astronauts out. It was a warm, moving experience, and the crew had a great time.

When we got to Chicago, Mayor Daley met us downtown, and we really had a parade. They had bands, majorettes, and they even had the fire boats out shooting different colored streams of water. That evening we spoke to the Chicanos, the Mexican-Americans, who were having a fiesta. Mayor Daley had suggested that we speak in Spanish, so they had something for each of us to say on a little card. We were so busy that I didn't have time to look at mine. Dave had actually memorized his card, and he expected Al and me to know ours by heart, too. It was an exhausting day, and I wasn't about to learn Spanish. Dave went through his spiel; Al said, "Olé," or something like that. As for me, I think I goofed.

I was beginning to develop the capacity for speaking extemporaneously for the first time in my life. I think having been to the moon left us less intimidated by this sort of thing, or maybe it was the fact that we were exposed to so many occasions when we were required to speak. As a midshipman at Annapolis I had been asked to speak without notes, and I refused because I couldn't do it. Now, obviously, I'd changed.

As I reconsider the whole experience and remember those extensive years of preparation, some things jell for me. The physical training was completely adequate. We had a few problems, but our procedures helped us solve them. I can't imagine any better preparation than the astronaut training program for the technical problems of the flight. The only thing I wasn't prepared for was the spiritual impact of the voyage to the moon. This sensing of God's presence, and the overwhelming feeling that He was there with us, was something I could not realize until some time after the flight.

As Mary watched me try to sort this experience out, she says she felt very strongly that something was happening inside of me. My first interview with *The New York Times,* which was scheduled before we were through debriefing, was the first opportunity I had to try to understand my own reaction to the flight. I had been so busy, so programmed, so totally absorbed with the challenge of the flight—that is, the scientific flight—that I had not yet been able to examine my feelings. I think Mary was more aware of the progressive change in me than I was.

There was nothing mysterious about one problem we brought back from the moon—the envelopes. This was the most controversial development of the Apollo program, and, although most astronauts were involved to some degree, NASA made an example of the Apollo 15 flight. It all started with 400 unauthorized postal covers or envelopes that our crew took aboard. (We actually had about 650 envelopes in all, but more than 200 were listed on the manifest.) On our third day on the moon, we played post office and cancelled the first stamps of a new issue commemorating United States achievements in space. With our own cancellation device—which worked in a vacuum—we imprinted August 2, 1971, which was the first day of issue.

The reprimand to our crew was the second ever issued by NASA. Curiously enough, the first went to Astronaut John Young for taking an unauthorized corned beef sandwich with him on the Gemini 3 mission, launched in 1965. As the spacecraft *Molly Brown* circled the earth, Young pulled out this hearty sandwich before the eyes of his startled buddy. I do not know whether John ate his sandwich or not. We certainly had to "eat" our envelopes.

I've said that of the 650 envelopes we had with us, 400 were not authorized. Of these, 100 were destined for an acquaintance in West Germany, Horst (Walter) Eiermann, formerly a factory representative for space products, and known to the astronauts at Cape Kennedy. We split the remaining unauthorized envelopes, Dave, Al, and I each taking 100. I had never thought about selling mine; I hadn't decided what I would do with them, and we had no agreement with each other about them.

Walter Eiermann contacted Dave first, and then Al and I were brought in on the deal. The initial contact was back in May of 1971, and then we had two additional meetings with Walter. Our agreement was that the envelopes would not be sold until the Apollo program was over. It was our plan to use the $8,000 each that Dave and Al and I expected to set up a trust fund for the education of our children. This was to be in the distant future, so the sales could in no way discredit the program. Dave carried the unauthorized 400 envelopes aboard the spacecraft in a pocket of his flight suit. The envelopes were onionskin and pressure-packed into a small and manageable parcel. You remember that we all three signed each one of the total 650 envelopes during the flight from Hawaii.

Dave mailed the 100 envelopes to Walter Eiermann in September, according to the agreement. We got a bankbook from Germany within a few weeks. Sometime in October, Dave came to me and said, "Jim, we are in trouble now— they are starting to sell the envelopes over there." We promptly sent the bankbooks back to Walter Eiermann, notifying him that he had broken the terms of the agreement and the agreement was off. We said that we didn't want to have anything to do with the deal. I don't know how Dave found out about the sale, but word had evidently leaked out in the European press. It was a tremendous blow to find out that we were being scuttled, actually torpedoed. Here we were, right on the eve of going over to Brussels on a goodwill trip, and the envelopes were actually being sold in Europe. This development did not make the flight over any more comfortable.

We were guests of the government of Belgium, and official guests of King Baudouin. The King was a fighter pilot and speaks English very well, so we had a marvelous time. We were guests at a formal dinner at the palace and then we showed a film, *Highlights of Apollo 15,* to the royal family and close friends. The King presented us with the Order of Leopold, and we were all relaxed and everything was lovely.

We had been invited to meet with scientists from all over Europe and had a slide presentation for this occasion. Well, it almost turned into a fiasco. We had all the slides in a carrousel—and a man who didn't speak English projected them for us. I think he started from the wrong end, and he had them in backwards. Dave shouted from the stage up to the projection booth but could not get his message across. Finally, he said, "Al, you go up there and get those slides fixed." When we got the show on the road, it went all right.

Since Apollo 15 had been advertised as the first extended scientific mission to the moon, Dave wanted to establish us as scientists. I had suggested we report our trip as pilots. Not only was I reluctant to pose as a scientist, I was becoming increasingly frustrated by being confined to the scientific message. I was much more aware of my religious or spiritual awakening, and I wanted to share this with our audiences.

But I didn't want to get NASA involved, because their policy is to stay out of this area.

Quietly and unofficially, I began moving out every weekend to visit churches. Then on October 27, 1971, I spoke before a crowd of 50,000 Southern Baptists at the Astrodome in Houston. I told them how God had changed my life. This was my first real testimony.

There was an explosion of interest. Word went out like wildfire that Jim Irwin had a spiritual message that he was willing to share with people, and tons of requests started coming in. This created a delicate problem of balancing my new mission with demands from the Space Center.

I was now assigned as backup crew for Apollo 17, and every minute in Houston was supposed to be spent training for 17. But being backup for 17 did not inspire me at all. The routine of going through that training for a third time, of preparing to take the place of a prime crewman, was the most unattractive thing I could imagine. I was faced with a good six months of intensive training, leaving the family again—all for a job I would never do, for I was backing up Jack Schmidt. Even if he broke a leg, I know they would delay the flight until it healed, so intent were they on getting a scientist on the moon.

Periodically, we got unsettling reports from Europe that additional envelopes were being sold for $1,500 each. It was amazing that things remained quiet at NASA for as long as they did. While the fuse on this powder keg burned down, my life was going through a process of total change. From the sort of person who had never been active in church, never cared to speak in public, I was in the new role of being totally active, of volunteering to speak as often as it was physically possible. Nor was I just speaking; increasingly I was inviting people to accept Jesus Christ as personal Savior.

It was one thing to tell them about the moon or the scientific mission, and then tell them about my spiritual voyage, and quite another to invite the audience to come forward and make a decision for Jesus Christ. It was a challenge to present the gospel in such a way that people would respond. I still haven't completely mastered it, and in the beginning the experience of trying to break through and get a person to accept Christ was filled with anxieties and frustrations.

Almost without realizing it, I began to cover a much broader territory: I was going down to Florida, out to California, up to Chicago, and across the country to Boston. Normally, I would rent an airplane and fly the whole family with me. Before I agreed to speak, I would explain that there would be expenses and what they were. I'd tell them that I always took my family with me, and that it cost about as much to rent a plane as it would to fly commercial. My average honorarium was about $200 or $300, and all checks were made out to The Jim Irwin Missionary Fund. Usually, I would rent a Cessna 320, a six-seater, and I'd let the children take turns helping me fly on the other control. I love flying, and the family had gotten so they enjoyed it.

There was a tremendous pressure on the weekends. More and more I found myself living in the future, after retirement, when there would be time to share the spiritual message. But I thought it was vital that I keep my credentials as a layman, that I have a real identification with the business world. A minister turns so many people off. I wanted flying to be a part of my life, and I wanted a job that would leave time for my mission and help me extend it. So, during the spring I looked around.

I was checking out companies that had fleets of planes or helicopters and might need pilots. Then I ran into Wally Shirra, and he put me in touch with Johns-Manville in

Denver. Paul Burke, their director of transportation, told me
they weren't hiring pilots, they were letting them go. But it
really caught Paul's interest when I told him that I wanted a
position that would allow me time to share my experience.
"Talk to our president, Dick Goodwin," he said. "Tell him
what you told me."

I told Dick how I felt about sharing my message and
about the people and the earth, and he showed great interest. He said, "Jim, why don't you just come to work for
Johns-Manville? You do this thing you want to do as much
as you want to do it, and when you have a few days free,
you spend them with us." This really was the answer to
prayer. I went to work for them on that basis.

In the spring of 1972, I was scheduled to be the Cap-
Com for Apollo 16, and Deke and I planned to fly down in a
38 together. We hadn't been together for a while, and when
I got with him he seemed preoccupied. He wasn't very talk-
ative that evening, but then he never is. Finally, he got
around to querying me casually about the envelopes and
asked how many we had carried. I told him; he gave me
another number, and I said it was not accurate.

"Deke," I said, "I'd rather not talk to you about it. You
should talk to Dave. He has the complete story. He handled
this whole thing." He said okay, he would.

The next day Deke called a meeting of Dave and Al
and me—but I was down about thirty miles south, fishing
with my old buddy Eddie Block, and didn't get the word, so
when Deke talked to Dave and Al, unfortunately I was not
present. When Deke got the complete story, he blew his top.
We were his guys who worked for him; he was like our
father. He saw this as a basic violation, a violation of trust,
so he got on the line and told his superiors what had
happened. Finally the story was given to the press.

This invited the response of Congress, and the Senate Space Committee felt that they had to investigate. Dave and Al and I were invited to Washington for a hearing in August.

Deke dropped by my office sometime in May, after Apollo 16 had splashed down. He said, "Jim, what are your plans after 17?"

I said, "Deke, I would like to retire. I've got other work that I would like to do, and I'd like to retire as soon as you can spare me."

"Why don't you start laying the groundwork to leave this summer?" he said. Al and Dave were on a trip, and they didn't know anything about this conversation. We three were still backup for 17.

Dave wanted to know what I had told Deke. I said, "I told him I'd like to retire."

"I don't think you should have done that, Jim," he said. He could see that the crew was being broken up.

I had always planned to retire when my twenty years were up. I had reached this point when I was prime crew of Apollo 15 and just about ready to launch, so I knew I had to delay retirement. Now there was nothing to keep me except my responsibilities for 17.

Mary saw the push to retirement as providence. She grieved with me over the envelope thing, but her response was clear. "It was just like the hand of God, giving you the push," she said. "He is saying, 'That's all, Charley. Get out and do *My* thing for *Me*.'"

That summer I went to the Southern Baptist Convention in Philadelphia and told the leaders of my denomination what I wanted to do. They offered to help me in every possible way. Out of our conversations came the idea that developed into High Flight. The objective was to create a nonprofit missionary foundation that could support and direct my efforts to share my message with people everywhere. My pastor, Bill Rittenhouse, helped me work out the design of this little corporation, and I named it High Flight after the title of the poem by John Gillespie McGee, a young Canadian aviator who was killed in World War II.

I had great support from Bill Rittenhouse, and we both worked very closely with the Southern Baptist Foreign Mission Board. In the late summer Brother Bill resigned from Nassau Bay Baptist Church after seven years as pastor in the Space Center community and became vice-president of High Flight. In the late summer of 1972, life crashed in on us and everything happened at once. The months ahead were a "high flight" in the sort of world ministry that I could not possibly have envisioned. But there was plenty of turbulence in the blast-off.

Back in June, after I had announced my intention of resigning from the Air Force, I made a public statement about my position on the envelope matter. I told the Baptist press, "The National Aeronautics and Space Administration had no choice but to reprimand us." I said we had acted in haste and under terrific pressure from the preflight and post-flight schedule, but this didn't excuse us. I said, "I don't think my mistake will damage my ministry through High Flight. It portrays me as a human, subject to human frailty. I hope it will open up opportunities for me to relate meaningfully to others who also have made human mistakes and need God's love and forgiveness." I also pointed out that we had had a change of heart and had refused the money eight months before the envelope incident was reported to the public.

The last day in July, when I was supposed to get my retirement papers, the sergeant rushed them over to me and in effect said, "Take them and run, before the Air Force changes its mind." I don't know whether that comment had any meaning other than underlining the fact that I would appear before the Senate Space Committee within the next few days. The atmosphere was charged with apprehension, and there was a sense of urgency about getting me out of there quickly, before anything happened.

Al and I flew up to Washington in a T-38 and compared notes on the way. We agreed that we three should share the responsibility equally. We were in this thing together, and we would take the blame equally; we didn't want to be divided at this point. Then Al and I thought about the irony of our situation. We had been back from the moon for less than a year, and during this brief period we had addressed a joint session of Congress as heroes, and now we were going back before these same senators in disgrace, because of this envelope scandal.

Among the Senators on the Space Committee were Clinton Anderson from New Mexico, Margaret Chase Smith from Maine, and Stuart Symington from Missouri. It was very formal. We were sworn in immediately and were the first to testify. Dave covered the envelope situation in terms of the whole inventory of articles that we took to the moon; he told about the extra watches and everything. Al told about his envelopes, and there were a couple of questions for me. They asked about the envelopes I had had with shamrocks, and the cancelled envelopes I had given away to personnel who worked in the mailroom down at the Cape; they couldn't understand how I could have given these valuable envelopes away to the help.

Deke had quietly dropped Dave and Al from backup as soon as I had retired, since there was no need to have the Apollo 17 backup crew involved in Senate hearings. Up to that point, Dave had enjoyed an untarnished reputation, and he must have suffered every sort of agony throughout the envelope episode. Sometimes I disagreed with Dave during the hearing, simply because we had a different memory of the facts, but we all three made absolutely full disclosures.

My new life really began in September; it was a fantastic month, and Mary and the children were with me most of the time. In that month alone we visited Oklahoma, Mississippi, Texas, Georgia, North Carolina, Wyoming, Minnesota, and probably other states that I can't remember. People everywhere were interested in seeing a moon man and in hearing what he had to say, both about the technical flight and the spiritual flight. And there was always this question: After you've been to the moon, what do you do? "Jim, what can you possibly do now?" people would ask. "Don't you feel let down, now that you are back on the earth?"

I'd tell them, "I've never had the letdown. The Lord has kept me on one great big high."

In the middle of September, Paul Stevens asked me if I would work with the Radio and TV Commission of the Southern Baptist Convention. They began training me to be a broadcaster, and I started doing radio and TV spots, announcements and all sorts of programs for them. They contracted for forty days of work during the year, and this helped us financially. Not only this, it helped me learn to relax in the environment of media, press conferences, and that sort of thing. I realized that you couldn't talk to everybody man to man; you have to rely on electronic communication.

During October and November our flights were higher and longer. On Friday, October 13, Mary and Joy and I flew out of San Francisco, across the International Date Line, to Japan, the next day arriving at Tokyo International Airport.

I spoke to Baptist churches and schools in several cities and had a brief meeting with Prime Minister Tanaka.

On October 18, I had an interview with Cabinet Minister Nakasone on a Tokyo TV program. Nakasone was Minister for Science and Technology, but the remarkable thing was that the interpreter was also a graduate of East High in Salt Lake City, class of '22. We were fellow alumni. Fantastic.

Just as we left for Korea on October 18, we heard that martial law had been declared. We landed at Kimpo Airport in Seoul on a beautiful clear day and were met by young ladies who presented flowers to all of us. I did not see President Pak in Seoul that afternoon because of the martial law, but I did present a Korean flag that I had taken to the moon to Prime Minister Chong Pil Kim. One of the highlights of my trip was my visit that night to the Air Force Academy on the outskirts of Seoul. The Commander, General Yoon, had been at Moody Air Force Base when I was, in 1954, and we had a tremendous conversation; he took his own Air Force wings off his uniform and pinned them on me.

I spoke to all the cadets who were assembled in the main auditorium. After making a speech on Apollo 15 and flying, I took the opportunity to give them my personal testimony. At the conclusion of the service, I gave the invitation to step forward and accept Jesus Christ as personal Lord and Savior. It was an overwhelming experience for me—fifty cadets came forward. I received them, shook hands with them, and prayed with them. This was the most moving group of young men I had ever seen.

Meanwhile Mary had visited some orphanages in Seoul, and during the tour she found a baby boy that she lost her heart to. We made special arrangements to go back to the hospital, but after a series of delays at the gate, and on the ward, we discovered that we were too late. Somebody had moved him away that afternoon. Mary's interest had been sort of tentative, just exploratory, but we were disappointed. We had talked many times before about adopting a son—you know we have three daughters and one son, and Jimmy would love a little brother. I guess if I had never gone to the moon I would never have thought of adopting an Oriental child. But my consciousness had been raised; those people had had so much tragedy, and I'd like to help. We decided that if we found the right little boy over there we would try to bring him back.

Korea really moved both Mary and me in the same way. It was a fantastic experience. This is really an electric country. As Mary says, God is moving tremendously in Korea; He is turning it upside down. We left Korea full of the beauty of the country and the responsiveness of the people, full of gratitude for the warmth of our reception.

We arrived in Taiwan on October 22, and the stay was rewarding but strenuous from a digestive point of view. One night we had a big rally in Sun Yat-sen Memorial Hall in Taipei. There were 5,000 in the audience, with the floor and the aisles packed absolutely to capacity. I had a terrible stomachache, and I asked the Lord to give me the right words and to give me strength. He really answered my prayer. About a hundred people came forward, and I felt that we had enjoyed a great experience together. The next day we had an opportunity to meet with Madame Chiang. She was most gracious and told me a Chinese myth about rabbits on the moon. I was able to give her a first-hand report: I didn't see any rabbits when I was up there.

Mary was thrilled to meet Madame Chiang, who was very warm and easy with her and asked how she had felt when I was making the flight. Mary said she had not worried because she had learned to take her problems to the foot of the Cross—not to lend them to God but to give them to Him. And Madame Chiang said, "Yes, in our prayer meeting group last Wednesday, we talked about this very thing." We spent about half an hour with Madame Chiang, and then we had to leave because her prayer group was coming in.

When we got to Vietnam, we could see the lights coming on in Saigon as we approached for a landing. We were briefed at the airport on what had been scheduled, and it turned out that we were going to have the opportunity to fly to Da Nang and also up to Hue near the DMZ. Mary decided that she would share the risk of traveling north, but we left Joy in Saigon with our friends, the Jim Humphries. After giving a brief speech and an invitation at the Base Chapel at the airport, we went into Saigon to the Caravel Hotel for the night.

After visiting briefly with the Mayor of Saigon the next morning, we had a visit to President Thieu scheduled for 11:30. We went through really tight security to the very sumptuous palace. We were met at the door by a presidential aide and some military men, who ushered us into a reception room where we were to await the President. Thieu is a small man with a round face who greeted us with almost boyish glee—he was very warm and very open. There was no hurry, no tension at all; this was one of the most relaxed visits we had on the trip.

I told President Thieu about my experience on the moon, and he talked about the great mountaintop experiences he had had. He showed us a prized possession, a piece of a moon rock that President Nixon had sent him. He had the enthusiasm of a young boy as he showed us his souvenirs from space.

At noon that day, we went back out to Tan Son Nhut

Airport to meet the Army crew who were standing by to fly us to Da Nang. It was a beautiful flight along the coast, following a chain of high mountains that reminded us of the California coastline south of Monterey. We landed in Da Nang and then got into a helicopter for the flight to Hue, where I spoke to a big group in a dark auditorium. Ninety-nine percent of the people were non-Christians, but I got a show of about fifty hands from those accepting Christ.

That night we flew back to Da Nang, where I was the guest of Lewis Meyers, one of the most impressive missionaries we met on the trip. I talked to a group of people from the whole community who had filled up Lewis' church, and then I spoke to the leaders of city government in the Chamber of Commerce Building. We had dinner with a group of community leaders—the best dinner we had on the entire trip. I remember that we had lobster, a meat dish prepared in the French way, and the most terrific French bread.

When we got back to Saigon the next day, I had a meeting with Ambassador Ellsworth Bunker. While I was busy with my rounds, Mary and Joy had visited an orphan home and found a little boy that Mary really liked; she said he looked like a football player. I did not have a chance to see him, but I told Mary that if she was convinced that was good enough for me. Then and there we began to make plans for our new little son, Joe, to join us in America as soon as possible. Our confidence could have been premature, but we had had the most enthusiastic and positive response from the adoption agencies in Saigon.

November 1, we flew on to Manila, and from there to New Zealand and Australia. I was delighted to get to Australia because of the boys at Honeysuckle Creek. This had been our prime tracking station, and with that big 210-foot dish they had gotten a great radio signal to us during the flight. A couple of astronauts had been by Honeysuckle, but nobody who had been on the moon, so the boys were delighted to see us. I dropped off a signed picture with a note of thanks for their help.

Mary was playing a more and more important part in my mission. While we were in Australia, I was scheduled to speak in the city hall in Sydney. The auditorium was completely filled when we got there, and hundreds of people were standing outside. We found there was a Baptist Church about two blocks away that was available. So Mary took the overflow crowd down to the church and ran the service herself. She did a great job.

The trip to the East had been great—an exhausting but tremendous flight for me. I really did feel at home everywhere in the world. People everywhere were willing to accept the message of my spiritual voyage.

I have a sort of clarifying sense of uncovering God's plan for my new life here on earth. God has permitted me to live on top of the mountains, and when I have been cast down, he has brought me back up on the tops of the hills again. Being able to give my testimony before people without fear, to share my spiritual voyage and ask them to accept Jesus Christ, has been a tremendous experience for me.

When I ask people to come forward, I am praying that God will touch every one of them. During my travels around the world, when I spoke to a group, I would visualize the earth as a sphere about the size of a basketball. In my mind's eye I locate myself—just exactly where I was on the earth at that moment with that audience. One time when I was praying this way, I could see a great hand in space, and it was directing a light onto the earth right at the spot where I was at that time. So now when I ask people to accept Jesus Christ, I ask for God's Spirit to give these people a new light in their lives.

The year 1972 ended with a most perfect trip to the Holy Land. Mary and the children were with me, and this was the most meaningful celebration of Christmas we have ever had. The Mayor made me an honorary citizen of Bethlehem, and I asked him if it would be possible to go to the Church of the Nativity. After elaborate arrangements, our family and Bill Rittenhouse's family were taken with an escort to the traditional birthplace of Christ on Christmas Eve. We all stood together in silent prayer in this holy place.

On Christmas day, I took part in a service planned for our group at the Garden Tomb. I ended my contribution by reading the poem "High Flight." The Tomb overwhelmed Mary and she wept like a child. After lunch that day, we drove to the mountain where Jesus delivered the Sermon on the Mount. That night I participated in a program with a group of American choirs, and after giving my testimony, I heard a performance of the Messiah. It was a miraculous season for us, and an incredible Christmas day.

Two days later, I had a meeting with Golda Meir, the Prime Minister of Israel. Mrs. Meir asked me if the flight to the moon had really changed my life or if it had merely strengthened my faith.

"Before the flight, I was really not a religious man," I said. "I believed in God, but I really had nothing to share. But when I came back from the moon, I felt so strongly that I had something that I wanted to share with others, that I established High Flight in order to tell all men everywhere that God is alive, not only on earth but also on the moon."

I told her that I thought my experience with God on the

moon was an answer to the prayers of millions of people who prayed for me during the voyage. Mrs. Meir turned to Mary during the course of the conversation and said, "I think this lady was as brave as you, sitting down on earth."

We pushed on from Israel to Jordan by bus, crossing the Allenby Bridge into Amman, Jordan, where we were to be the guests of King Hussein. I had a most delightful visit—the King and I had long conversations about flying, and about the inspiration of the sky and the moon. The King's brother, Prince Hassan, took me out in his Falcon Jet to Hassan Air Base. Then we went out to the flight line to the alert hangars, and they flew a couple of 104s for me.

The 104s scrambled very quickly with the afterburner takeoffs and climbed right up, and my heart really thrilled. Then they swung around and made several low passes over the base—I thought, enviously, I would have been grounded if I had done that. It was really moving. I wondered if I had done the right thing in retiring from the Air Force.

The whole experience in the Holy Land was the most inspiring way to turn the corner of the old year into a bright new year for High Flight. The experience moved us and gave us a chance to rededicate ourselves to each other as a family and to our mission. Halfway around the world, hearts had been opened to us. We were sharing something much deeper than science, and the response helped us feel that our mission was blessed.

Mary was a full-fledged partner now; she was sharing the flight with me, and so were the children. And at that point, I couldn't help but look ahead and think of little Joe joining us back in Colorado Springs. He would be a part of our lives and a part of our family worship service that, above all things, had really brought us together. Being in the Holy

Land gave a sort of warm promise, a glow to the future. I am glad that I could not see everything that lay ahead, but I was right in being hopeful about the future.

When we got back to the United States the pace never slackened. We were looking forward to the arrival of Joe, but there were many delays because of the problems our country was having with South Vietnam. The flight was set up in February, but at the last minute it was cancelled again. Finally Joe arrived in March, and we all had the feeling that our family was complete.

Mary is a lot closer to me now than she has been—we have both changed a great deal, and she is tremendously reinforcing to me. She was telling me earlier this year that I was going to have to learn to pace myself, to come to grips with myself. She said she was convinced that it took drastic measures to get through to me. Only the Lord knows what's right, and He is going to show you, she said. Well, He did.

On April 4 I was playing handball in Denver, and I had a heart attack. At that moment, my high flight was brought down again. I was grounded one more time, or, as Mary said, "It was God taking charge again." Somehow my Mission Director was changing my flight plan. I know that He has a plan for my life. It's my business just to wait on Him, and He will give me my orders.

EPILOGUE

I WAS SURPRISED by God. I couldn't imagine that I, Jim Irwin, would have a heart attack. This was a hard thing to accept when it first happened, and when I had a chance to reflect on it, it was embarrassing to me. This heart attack made me more dependent on the Lord; it made me accept my own human weakness. I wonder if accepting weakness is not the first step toward new spiritual power?

I was never frightened, and I remember having the thought that if the Lord was calling me, I was ready to go. It was a clear, definite feeling of surrender. I had been up and down so many times; I was passing through another valley, another period of darkness. That first night after I had the heart attack was probably the most comfortable night of that week. They had me sedated so I was vague, but I was completely at peace with the world. I remember wondering, very quietly, why did this happen?

That first night I didn't articulate any prayers; I formed them in my mind. I prayed that Mary, the kids, and all of my family would be able to accept my death, if it happened. I had accepted this. I did not pray for my own recovery until after I left the intensive care unit. At first I was surprised that God had given me another night, and as I began to understand that God ruled the night, the morning came—God had

245

a plan for me, He had brought me into another night so that I could have another day. It is not a physical force; it is a spiritual force that helps us through the night. This time when I passed through the night, I had no fear; I was ready to accept anything, because I had Jesus Christ there to strengthen me.

You might guess that one of my immediate concerns was about flying. When I was in intensive care before anybody knew whether I'd live or not, I thought, how long am I going to be grounded this time? I don't think it was until I got down to the regular ward that I got an answer to that. The doctor told me that I'd had a myocardial infarction, a serious heart attack, and I would be grounded from flying as a pilot for two years. This means that High Flight is grounded. To do what I have been doing in High Flight means that I have to fly, and if I can't fly, I'll have to have someone else fly me. This means that I will need help, and I need help for more than flying. I had been sustaining High Flight with my honorariums for speaking. I had been putting money in, and I had been putting most of my time in. I had gotten to feel that I was High Flight. Well, it is obvious as a result of the heart attack that now it's really up to the Lord to decide whether High Flight is going to continue or not. I'm not going to ever be able to do it all.

It has been abundantly clear that we have got to have other people join our flight, full time or part time. We are bringing POWs in who have come back with great depths of spiritual understanding and want to share this with people around the earth. We are encouraging astronauts to join us after they leave the program if they have a spiritual message that they want to share. I really see the hand of Providence in what happened to me. I believe the Lord sustained me to a certain point; then He decided that I had to get a rest and reflect on what my future and the future of High Flight should be.

Well, my first reflection was that a heart attack was the only way to slow me down. Looking back on my schedule, it had reached the point during the last few weeks that I was dragging. I dreaded the commitments that I had made, every day on the calendar filled in, edited, re-edited—for months ahead. If you look at the calendar you can see a prescription for disaster. I had wondered how I was ever going to sustain myself until the month in Western Europe, and how was I going to make it through the quick turnaround and then the trip to Eastern Europe. I was getting to be resentful of the schedule, of what it was doing to my relationships with my family. I had hardly had time to get acquainted with little Joe. Mary believes that the Lord wanted me to slow down.

Ironically, a little over a week before the heart attack, I was down at NASA for my annual physical. It was an exhausting thing—took almost half a day. They told me I was in better shape than I was a year ago. (I actually weighed one pound less than the previous year.) I knew it wasn't completely true, but I had at least satisfied NASA. So I breathed a sigh of relief and kind of celebrated. But the next week was really a rough one.

On Saturday, March 31, I flew out of Colorado Springs to Cleveland. I was driven from there to Amherst, where I spoke that night. From that afternoon on, it was kind of a blur. I didn't have the right clothes for a reception they had at the country club. Immediately afterwards I got some really irritating questions at a press conference. "How much are they paying you to come here to speak? Are you making more money now than you did as an astronaut?" I didn't have time to take a shower before I had to speak at the Chamber of Commerce banquet in Amherst that night. The next morning

I took part in two services at the Church of the Open Door in Elyria, Ohio. I even launched a real, live Estes rocket, before I left on a 2:00 P.M. flight for Atlanta.

It was Atlanta to Macon, Georgia, with police escort, and no time to eat. I preached over TV from the Mabel White Memorial Church in Macon; signed autographs in a jostling group for an hour; ate at 10:00 P.M.; it didn't sit well and I didn't sleep well; had breakfast with Baptist ministers at 7:30 A.M. the next day and spoke; drove to La Grange with help from state troopers to speak at the La Grange College convocation; quick lunch; drove back to Atlanta for 2:00 P.M. speech at Westminster School. From then on it was evening service at the Sylvan Hills Church; autographing, handshaking; the next morning radio and television followed by fundraising downtown at noon; back to Colorado Springs in the late afternoon. Mary met me and we talked about the written exam I was going to take for the Air Transport Rating the next day in Denver.

Well, on the way to Denver, a little after 7 the next morning, I was driving over icy roads and studying with my papers on the steering wheel. I slid off the road, and finally with a little help from a Conoco wrecker, I got into Denver, late, for a five-hour exam. I worked right through lunch, and then rechecked the sixty problems and solutions. Finally, I got out on the handball court at Lowery Air Force Base; that is, after I had lifted a few weights, run the track, done a few dips on the parallel bars.

I was playing cutthroat with Dick Sauer and Lee Hollingsworth, my Lear Jet instructors. All three of us were on the court. Within two or three points, I felt this pressure right in the middle of my chest. It was a force pressing down, like a hand pushing in right there. I had never felt this before.

EPILOGUE

I played up to five points, but it wasn't going away—it was getting worse. It was most embarrassing. I excused myself.

I had already started sweating; now it just ran off me. The pressure was persistent, and by this time my arms were going numb. I lay down on a bench; I was struggling to breathe, struggling to get oxygen, and my arms were tingling like they'd gone to sleep. Then suddenly I got the dry heaves. As I ran into the men's room I told the attendant to call an ambulance.

As it turned out, the ambulance was busy, so I got a modified jeep—open in the back. I sweated and froze and shook the ten minutes to Lowry, and then another ten minutes to Fitzsimmons Army Hospital, where a quick-witted doctor at Lowry had sent me. I was totally conscious, trying to breathe—moaning when they got me into the emergency room at Fitzsimmons. The guys were all over me—sticking needles into me, and then searching around with them for veins. One tried to rip off the top of my new tennis suit; I pushed him away, sat up, and took it off. That really made me mad. Then they pushed me back down and started again. Within minutes, they had gotten a sedative in me and I was up in the intensive care unit.

During the first week in intensive care, I was very hazy. They were putting sensors on me, and I felt like I was being sensored for a flight. They had needles in both arms, both legs, in veins in both feet, and I had tubes up my nose, through my mouth, everywhere. Then, before I knew it, I had this bottle in my hand, and they were all telling me I must urinate. The lights were on, the doors open, there were nurses everywhere, and they told me they were going to catheterize me if I didn't urinate. It was just like trying to urinate on the launch pad—but this time it was harder. Only

249

when they finally turned off the lights and vacated the room did I succeed.

Mary came up that evening, and she stayed in an adjoining room most of the critical first three days. She said that she was absolutely confident that God had a plan for me, and that I was going to survive. After a week or so, I began to pray that, if it were God's will, I would recover. The hospital staff must have believed that I was recovering, because soon all the doctors were coming in to meet me. The fact that I am an astronaut made it sort of a sideshow—although I was in intensive care, I was receiving visitors and signing autographs. They were also a little apprehensive about having a celebrity dying on their hands, so they gave me a tremendous amount of attention—maybe too much.

But I was able to survive the hospital, and when I got home, it was just wonderful. Mary, you know, is the greatest nurse I could ever have asked for. And she took care of me the way she did ten years before, after the crash. It has been an ideal relationship since I got back—it has been a vacation; she has been spoiling me. It is almost like a honeymoon. Our old differences have paled. It is happening very simply, because when you know Jesus Christ, differences fade away. Mary is a full-fledged partner now in High Flight. She is a very strong force. She is witnessing and speaking; she even gives the invitation, and she must say the right thing, because people come forward to accept Christ.

Mary tells me that we have never really had any time alone before—that this is a new experience for us. And I see it as a new life within a new life. We know what quiet times together have been like—lying on the rug listening to music or playing chess. And I know she dreams of running through the leaves in the forest or, as she says, "just lying on a blanket in the high mountains and watching the clouds go by." So I know there is a tremendous future for Mary and me—we haven't begun to realize the potential of our relationship—and for the whole family. I'm convinced that God wants me to spend more time with my family.

After the heart attack my prayers were for Mary and the family, and then they were for myself, and finally they are for High Flight. I am a stubborn man, but now the Lord has my attention. I know I need help, and I am willing to accept it. I am grateful as the Lord reveals His plan for this flight, the Highest Flight of all.

James Benson Irwin, age 6 months, Pittsburgh, 1930.

Jim with younger brother, Charles, New Port Richey, Florida, 1941, the year he first accepted Jesus Christ.

Orlando, Florida, 1944; first dreams of flying.

Young pilot with Dad.

At test pilot school, Edwards Air Force Base, California, 1961.

Suited up for the important
LTA-8 testing, 1968.

Family atop the Vertical Assembly Building at Cape Kennedy, 1966.

Parasailing in Acapulco—always a new challenge.

With Dave Scott on geology trip, training for Apollo 15.

The Apollo 15 crew. Al Worden, Dave Scott, and Jim Irwin at the Air Force Academy in Colorado Springs with the falcon mascot.

Father and son astronauts, 1970.

(left) *Irwin salutes American flag on the moon,*
with Mount Hadley Delta in background, the
Falcon, center, and the Rover at right.

Irwin with lunar dune buggy (the Rover) on the moon;
St. George Crater (behind Irwin's head) about three miles away.

With Mom and Dad in 1972,
after the Apollo 15 Flight.

The family in 1972. The children from left:
Jimmy, nine; Joy, thirteen; Jill, eleven; and Jan, seven.

Showing Anita Bryant and Bob Green around the Manned Spacecraft Center, Houston, 1972.

With General Erik Wickberg of the Salvation Army.

Sharing the message of God.

*(left) Vice President Agnew presents
NASA Distinguished Service Award
to Colonel Irwin.*

*Crew visiting President Nixon at the
White House with Mrs. Scott, second from left,
and Mrs. Irwin, second from right.*

Jim Irwin presents Golda Meir with an Israeli flag that had been to the moon.

Sons Jimmy and Joe at the ranch in Colorado Springs, May, 1973.

48098